Trends and Innovations in Food Science

Edited by Yehia El-Samragy

Published in London, United Kingdom

Trends and Innovations in Food Science
http://dx.doi.org/10.5772/intechopen.101294
Edited by Yehia El-Samragy

Contributors
Maksud Hakim, Ali Syahban Amir, Raúl Alberto Reyes Villagrana, Ana Luisa Renteria-Monterrubio, América Chávez Martínez, Juliana Juárez-Moya, Rogelio Sánchez Vega, Juan Manuel Tirado, Zulqurnain Khan, Umar Farooq, Afshan Shafi, Aliza Batool, Zuzana Hlaváčová, Eva Ivanišová, Peter Hlaváč, Ľubomír Kubík, Daniela Kunecová, Monika Božiková, Vlasta Vozárová, Jofrey Raymond, Teresa Garde-Cerdan, Eva P. Pérez-Álvarez, Pilar Rubio-Bretón, Noelia López-Giral, Ankur Aggarwal, Tarun Verma, Sumangal Ghosh, Adelodun Kolapo, Abosede Fawole

First published in London, United Kingdom, 2022 by IntechOpen
IntechOpen is the global imprint of INTECHOPEN LIMITED, registered in England and Wales, registration number: 11086078, 5 Princes Gate Court, London, SW7 2QJ, United Kingdom

British Library Cataloguing-in-Publication Data
A catalogue record for this book is available from the British Library

Additional hard and PDF copies can be obtained from orders@intechopen.com

Trends and Innovations in Food Science
Edited by Yehia El-Samragy
p. cm.
Print ISBN 978-1-80356-065-6
Online ISBN 978-1-80356-066-3
eBook (PDF) ISBN 978-1-80356-067-0

Meet the editor

Dr. El-Samragy had over four decades of a professional career bridged between academia and industry. He was a Professor Emeritus of Food Science at Ain Sham University, Cairo, Egypt, and a Visiting Research Professor at Cornell University, Ithaca, NY and Utah State University, Logan, UT, USA. He was an International Expert Trainer in Food Safety and Quality Management Systems. He worked as an Expert at some international organizations including FAO, UNIDO, UNDP, JECFA, ISO, USAID, ACDI-VOCA, and DANIDA, in different projects of technology transfer, food standards, food product development, waste utilization, cleaner production, implementation of integrated management systems. He was IRCA Lead Auditor/Tutor of QMS, and Food Safety (HACCP & ISO/FSSC 22000) (IRCA Certificate # 01182132), and Lead Instructor, FSPCA Preventive Controls for Human Food Course (FSPCA Certificate # d16e213f) and FSPCA Foreign Supplier Verification Programs (FSPCA Certificate # d26bcf6b). Also, he registered and approved to deliver Food Safety and HACCP training and examinations leading to Highfield Qualifications (Highfield Tutor # 29012). He had extensive experience in delivering training courses on QMS, HACCP, and ISO/FSSC 22000 in Egypt, Libya, Sudan, Zambia, Tanzania, Ghana, Sierra Leone, Liberia, Gambia, South Africa, Uganda, Saudi Arabia, Yemen, Jordan, Dubai, Sharjah, Syria, Bahrain, Lebanon, Kazakhstan, Russia, USA and Canada.

Contents

Preface

This Edited Volume is a collection of reviewed and relevant research chapters, concerning the developments within the Trends and Innovations in Food Science field of study. The book includes scholarly contributions by various authors and is edited by a group of experts pertinent to agricultural and biological sciences. Each contribution comes as a separate chapter complete in itself but directly related to the book's topics and objectives.

Contents include: "The Function of a Coffee Shop as a Social Cultural Entity", "High-Intensity Ultrasound and Its Interaction with Foodstuff and Nanomaterials", "Nutrigenomics: Challenges and Opportunities", "Value-Added Foods: Characteristic, Benefits, and Physical Properties", "Context-Specific Food-Based Strategies for Improving Nutrition in Developing Countries", "Pulsed Electric Fields as a Green Pretreatment to Enhance Mass Transfer from Grapes of Bioactive Molecules: Aromatic, Phenolic, and Nitrogen Compounds," "Heavy Metal Residues in Milk and Milk Products and Their Detection Method", "Optimization of Cassava (*Manihot esculenta* Crantz.) Fermentation Processes for Food-Secured Twenty-First Century Africa". The target audience comprises scholars and specialists in the field.

IntechOpen

Note from the Publisher

It is with great sadness and regret that we inform the future readers of this book that the Editor, Prof. Yehia El-Samragy, passed away shortly before finishing the book and having a chance to see its publication.

Prof. Yehia El-Samragy was IntechOpen's dear collaborator and was an editor of 2 books, "Food Additive" and "Food Safety - Some Global Trends".

This collaboration continued until his final days when he was acting as the editor of the book "Trends and Innovations in Food Science".

We would like to acknowledge Prof. Yehia El-Samragy's contribution to open access scientific publishing, which he made during the years of dedicated work, and express our gratitude for his pleasant cooperation with us.

IntechOpen Book Department Team, October 2022

Chapter 1

The Function of a Coffee Shop as a Social Cultural Entity

Maksud Hakim and Ali Syahban Amir

Abstract

This book chapter aims to determine the process of changing the meaning of a coffee shop from traditional to modern, to find out the change in the meaning of a coffee shop as a space for socio-cultural entities, and to find out the change in the meaning of a coffee shop as an effective political tool. This book chapter uses a naturalistic (qualitative) paradigm approach to analyze and construct the change in the meaning of a coffee shop as a space for socio-cultural entities in the municipality of Makassar, South Sulawesi. The selection of this location is based on the consideration that Makassar Municipality is a large city that has a heterogeneous population. This book chapter shows that the change in the meaning of traditional coffee shops to modern coffee shops is characterized by simple facilities that develop into public spaces equipped with facilities and information such as television, newspapers, and internet networks to meet the needs and satisfaction of visitors. Coffee shops have become a socio-cultural entity. Coffee shops have been used as a means of political discussion and self-image, both formally and informally, socialization of work programs with political figures or authorities, and digging information effectively from visitors.

Keywords: coffee shop, socio-cultural entity

1. Introduction

In the mind of most people, drinking coffee is just to increase the body's resistance to sleepiness. In fact, coffee can provide so many benefits for human life. Ibn Sina (1000 AD) has investigated the chemistry of coffee. His document is the first known document dissecting coffee from medical and health sciences. The results of subsequent studies found that drinking coffee can prevent uterine cancer, and can improve memory and concentration for workers. Drinking coffee for children once a week is believed to strengthen the heart. Meanwhile, for women, drinking coffee can be used as an alternative choice in dieting, because the efficacy of coffee can delay hunger.

In Makassar city, before 2000, the existing coffee shops were only used to satisfy the desire for coffee. In the next decade, the function of the coffee shop became more diverse. Starting from a place to discuss daily social problems, a place for socialization for the rulers and politicians, a means of business meetings, to a means of self-expression.

The coffee shop has developed into a medium for expressing lifestyle and class identity. Visiting coffee shops has turned into a symbolic code for certain circles to actualize their existence in social groups. This can be seen from the shop's visitors, who are now not only men but also women.

Changes in coffee drinking patterns and behavior are increasingly developing along with globalization, which is emphasized by media imperialism. For Klapper, the mass media is able to manipulate consciousness. By Merton, the media can smooth out coercion so that it looks like persuasion [1]. The globalization of the media has played a role in the spread of the culture of enjoying coffee in coffee shops. Contemporary culture is related to people's tastes. According to Bourdiau, the taste is an opportunity both to experience and to assert one's position in the environment [2].

Visiting a coffee shop is not just enjoying a delicious cup of coffee. Now the attention is no longer on what is drunk in coffee shops, but more than that, namely how coffee is processed, served, and drunk. Changes in people's patterns and behavior in enjoying coffee cannot be separated from the influence of the globalization of the coffee shop business which has increased in recent years. The coffee shop business in the era of globalization does not only work on coffee products, it is further than that, namely consumers by formulating contemporary business jargon which is full of symbolism. Enjoying a cup of coffee in a modern coffee shop for some people has become a symbol of luxury, exclusivity, and prestige, as well as a marker or identity of social class. The change in the meaning of the coffee shop is also inseparable from the social interaction of the community.

In Blumer's perspective of symbolic interaction, the change in meaning can be seen from three premises, first, humans act towards something based on the meaning that something has for them; second, the meaning comes from one's social interactions with other people; and third, the meaning is perfected when the process of social interaction takes place [3].

As a result, it also has an impact on changes in society. Consumerism began to take root in urban society. Consumptive-ism in this case is defined as the use or consumption of goods that are pure because of the demands of social prestige and not out of necessity. This fact shows that cultural space is not only created by the community itself, but it is also a part of the expansion of global culture which intentionally or not has expanded to coffee shop spaces. Cultural globalization is a form of transnational expansion towards cultural grafting (heterogeneity). This process towards homogeneity is often associated with cultural imperialism which is characterized by increasing international influence on certain cultures. So, do not be surprised if coffee shops will one day replace social spaces that have been established before. The coffee shop has become a part that not only has become a consumptive medium for certain classes but has shown the socio-cultural characteristics of the community.

The proliferation of coffee shops in Makassar city has encouraged the growth of public spaces. Coffee shops are no longer just a means of meeting the need for coffee but have changed their shape and face as a social and cultural space, capable of becoming cultural glue. This is possible because coffee shops are able to reduce the barriers to differences in the social, economic, and cultural backgrounds of visitors, by designing an atmosphere and communication that is as loose and fluid as possible. So, that every citizen, from various social classes, can fuse together in a social space full of dynamics. Coffee shops can also trigger the creation of openness and exchange of information because the social space provided by coffee shops is so free and open. Coffee shop visitors are free to talk about problems in social and cultural life without intervention and intimidation from other groups or by the authorities and bureaucracy.

The phenomenon of the rapid development of coffee shops in Makassar city today, not only reinforces existing social changes but also serves as a buffer for socio-economic strength because coffee shops are able to accommodate the diversity of community members. In addition, coffee shops are not only able to increase the economic level of middle and upper-class coffee shop entrepreneurs, but also provide income for informal sector entrepreneurs who are generally lower-class people.

The change in the meaning of coffee shops in Makassar city from traditional coffee shops that focus only on coffee consumption as mere physical fulfillment, to modern coffee shops equipped with various facilities that allow coffee shops to play a role as a means of business, information, socialization, and imaging. These coffee shops with modern management are now more existent in Makassar city, while traditional coffee shops seem to be running in place and some of them have even gone out of business.

The habit of drinking coffee in the urban community of Makassar city is more dynamic with very loose communication patterns. The author's initial observations about the characteristics of coffee shop visitors in Makassar city, that there are several characteristics of coffee shop visitors. The first, are those who are classified as coffee connoisseurs to satisfy their physical needs. Second, are those who visit coffee shops because they enjoy socializing and building relationships. For people like this, with expenses that are not so big, they can hang out as much as they want in the coffee shop while enjoying the atmosphere provided at the coffee shop, including the sensation of serving, good-looking waiters, and free internet facilities. Third, are those who are classified as business people. They use the shop as a space for business transactions or use the coffee shop's social space to build networks and business relationships. Fourth, are those who use coffee shops as a means of obtaining information. Included in this group are students and journalists. Fifth, are political actors who make coffee shops a place to disseminate ideas and political imagery.

1.1 History of coffee and its distribution

The history of coffee has been recorded since the ninth century. Coffee is a plant native to Ethiopia which was originally only grown in highland areas. When the Arabs began to expand their trade, coffee beans have also expanded to North Africa. It was from North Africa that coffee beans began to spread from Asia to the European market.

Since the beginning, the coffee plant was not as popular as it is today. Coffee is just an ordinary plant with no known benefits. The beginning of knowing the benefits of coffee for humans is illustrated in the story of a goat herder in Ethiopia. The benefits of coffee were first discovered by Kaldi, a goat herder in Ethiopia. Kaldi found his goats behaving strangely after eating coffee plants. The goats were jumping around like they were drunk.

Kaldi was surprised and found out what was causing the goats to dance. Kaldi then saw a bunch of shiny red seeds in the bushes. The grain—later known as coffee—was being eaten by the goats. With curiosity, finally felt the difference after trying to eat the seeds.

Meanwhile in Indonesia, coffee was brought to Indonesia by a Dutchman named Zwaardecroon, who brought some seeds of the Arabica Mocca coffee plant from Mecca to Bogor in the seventeenth century. Arabica coffee was first grown in a place east of Jatinegara. The place is now known as Pondok Kopi. Arabica coffee then spread to various areas in West Java, such as Bogor, Sukabumi, Banten, and Priangan. From the island of Java, enter other areas such as the islands of Sumatra, Sulawesi, Bali, and Timor [4].

1.2 Theory of habitus and environment

Bourdieu formulated the concept of habitus as a sociological and philosophical analysis of human behavior. Habitus are values that are lived by humans and are created through a process of values that lasts a long time so that they settle into ways of thinking and patterns of behavior that persist in humans.

Dialectically, habitus is a product of the internalization of the structure of the social world [5]. Habitus is a subjective structure that is formed from the experience of individuals relating to other individuals in a network of objective structures that are in social space. Habitus can be said to be a cultural unconscious, namely history that is unconsciously considered natural, which is formed from the results of learning through parenting, playing activities, and also community education in a broad sense.

Bourdieu argues that the way actors feel is based on their position in the social space and constructs social life as a sociological study. However, perceptions and constructions that occur in social life are driven and controlled by structures. Bourdieu's thinking is reflected in his own definition of his theoretical perspective:

> *"Objective structure analysis cannot be separated from the analysis of the origin of the mental structure of individual actors which to some extent is the product of a combination of social structures and cannot be separated from the analysis of the origins of the social structure itself" [6].*

In the view of Bourdieu [5] habitus are:

> *Mental or cognitive structures are used by actors to deal with social life. Actors are provided with a series of internalized schemas or patterns that they use to perceive, understand, and be aware of their social world. It is through these patterns that actors produce their actions as well as judge them.*

So, the habitus will be different, depending on the form of a person's position in social life. Not everyone has the same habits, people who occupy the same position in social life tend to have the same habits. In this sense, habitus can also be a collective phenomenon.

Habitus allows people to understand the social world, but having multiple habitus means that social life and its structure cannot be imposed uniformly on all actors. Actions mediate habitus and social life. On the one hand, habitus is created through practice (action); on the other hand, habitus is an action created by social life [7].

Habitus merely suggests what people should think and what they should choose to do. According to Bourdieu, habitus functions below the level of consciousness and language, beyond the reach of observation and control by volition [8]. But it manifests itself in our very practical activities such as the way we eat, walk, talk and even blow our noses.

Bourdieu focuses on differences in the "taste" of beauty between various social classes towards various cultural objects that contain the value of beauty. For Bourdieu in Ritzer and Goodman [9]:

> *Taste turns out to be also a practice which, among other things, helps to give an individual or another an understanding of his or her position in the social order. Taste helps tell other people who have similar preferences helps differentiate them from others who have different tastes.*

In Bourdieu's view, "through the application of habitus and taste, people classify objects and at the same time they are in the process of classifying themselves". Bourdieu further explained that:

> *Taste is an opportunity both to experience and to affirm one's position in the environment. However, the social class environment has a big influence on people's ability to play this game; they are greater in their ability to defend their own tastes and oppose the tastes of people in lower classes [7].*

Bourdieu also connects taste with habitus. Tastes are shaped by long-lasting habitus; not shaped by superficial opinion and rhetoric. Furthermore, Bourdieu explained that the environment and habitus mutually determine each other. This can be seen from his explanation of the two things, namely:

> *A steady habitus is only formed, only functions, and is only valid in an environment, in relation to an environment the habitus itself is "the environment of existing forces", a dynamic situation in which power is only incarnated in relation to certain tendencies. This is why the same habitus gets opposite meanings and values in different environments, in different configurations, or in opposite sectors of the same environment [2].*

Bourdieu views that the driving force of all human behavior is the pursuit of honor. Bourdieu argues that his main goal is "to exist in a social space, occupy the point where he becomes an individual in a social space, is to become a differentiator in order to become prominent in that space [10]. Bourdieu further argues that the environment offers almost endless opportunities to pursue honor.

1.3 Theory of social interaction

According to Kaldun, "humans are basically created as social beings, namely creatures who always need other people to maintain their lives so that their life with society and social organizations is a must" [11]. Because the relationship between one another, both in the form of groups and individuals is a picture of the development of society in realizing its dynamism.

Symbolic interactionism theorist Cooley [12] sees "society is a mental phenomenon, the relationship between people's ideas". Meanwhile, Thomas, Henri, Susan, Bridget, added that "society is in my mind like the relationship and mutual influence in certain ideas which are named "I". Bahtiar [12] sees "society and individuals are not two separate realities, but two sides or aspects of one and the same reality. The two are like two sides of a coin that cannot be separated."

These human characteristics illustrate how meaningful social interaction is in human life. Young and Raymond stated that "social interaction is the key to all social life because, without social interaction, there will be no life together" [13]. Social interactions that take place are basically based on several factors, including imitation, suggestion, identification, and sympathy. These factors can move individually or in combination. For this reason, according to Soekanto [14], "social interaction is impossible if it does not meet two conditions, namely the existence of social contact (social contact) and communication". Meanwhile, Mead [15] is of the view that "only by adjusting to the expectations of others, interaction will be possible".

In sociology, the unit of analysis is a social event called social interaction between two or more people. Social interaction can be classified into various types of social relationships that are fostered by a number of people, actors from one or two groups/organizations. Various social relations between people from one or two groups called inter-group social relations according to their characteristics can be classified into various social processes, there are social processes that bring people closer and there are social processes that distance people.

Social interaction can be classified into several forms of interaction. Gillin and Gillin [16] divide the form of interaction into two parts, namely: (1) associative processes (accommodation, assimilation, and acculturation), and (2) dissociative processes (competition, opposition). Meanwhile, the form of interaction is divided into three parts, namely: (1) opposition (competition and opposition), (2) cooperation that results in accommodation, and (3) differentiation (each individual has the right and obligations on the basis of differences in age, sex, and work. As for Tomatsu Shibutani, distinguishes the form of interaction into (1) accommodation in routine situations, (2) expression of meetings and suggestions, and (3) strategic interaction in conflict; development of mass behavior.

In discussing social interaction for the Bugis-Makassar community, the tradition that should be put forward is tudang sipulung. The term tudang sipulung literally means "sitting together", which has the meaning and purpose to solve their needs and problems in their lives and daily lives in a democratic manner. A pallontara ("interpreter of Lontara"), Andi Burhanuddin explains in Bugis language that:

> *naiya riasenge tudang sipulung, iyanaritu mallari ade-e napogau toriolota'. Tudang maddepung-deppungeng, tudang mallewo-lewoang nasibawai akkatta maelo sipatangngareng nenniya maelo mala ada assimaturuseng.*
>
> *(What is meant by tudang sipulung is a tradition that is often carried out by the ancients (our elders). Sitting together, gathering with the aim of deliberation to reach consensus" [17].*

The presentation indicates that the Tudang sipulung tradition has been carried out by the Bugis-Makassar community as a common space for deliberation and consensus in order to find solutions to problems faced by the community.

Anwar Ibrahim [17], states that "all problems in people's lives can be blamed for being blamed". The implementation of tudang sipulung can be official or informal. Tudang sipulung, which is informal in nature, is usually carried out in a family environment or between families who discuss family issues such as marriage and proposals. While matters concerning social issues, or important decisions within a village, between villages or kingdoms, are usually carried out officially led by a matoa (who is elder according to custom) as the leader (king) of a village/country (wanua).

This illustrates that the position of tudang sipulung for the Bugis-Makassar community is very important. This can be seen from its goal of being a bridge or mediator for the resolution of various problems, both problems with a small scope, namely family, as well as issues with a wider scope, such as political and state issues. For the people of Makassar city who are more modern, the emergence of coffee shops today is not only a means of satisfying the need for the pleasure of coffee, but more than that, namely as a gathering place while discussing various life problems, both those concerning social, cultural and political issues. . So it is not an exaggeration to gather at a coffee shop to be described as a manifestation of the tudang sipulung tradition.

1.4 Theory of symbolic interaction

Communication is a means of driving a social process that allows interaction between humans and makes humans as social beings. Humans communicate with symbols. According to Rivers, Jensen and Peterson [18], "humans are creatures who know how to react to their physical environment, but also to the symbols they make themselves". Meanwhile, Kenneth Boulding compares humans with animals in terms of communication, according to him "animals do not have the notion of consciousness and a symbolic environment (language, art and myth) like humans do; So, humans are unique not only because of their ability to reason but also because of their symbolic world [18].

Humans always communicate with symbols. This is what underlies Rivers et al see humans as creators of symbols. According to him "the world is a pseudo, a network or a series of symbols of his creation". In line with Rivers et al., Epictetus argues that "what strikes humans are not objects, but their own opinions and images about these objects" [18]. In interacting, for Mead, "mind and self are part of human behavior, namely part of their interaction with other people. This interaction allows him to know the world and himself" [19], thus, "thinking can be understood as a result of internalizing the process of interacting with other people", continued Mead [15].

Meanwhile, another thinker of the flow of symbolic interactionism, namely Blumer emphasized that symbolic interaction includes "interpretation of action". Blumer rests on three premises, namely: "(1) humans act toward things based on the meanings that things have for them, (2) those meanings come from one's social interactions with other people, and (3) these meanings are perfected when the process of social interaction takes place" [20].

Symbolic interactionist theorists tend to agree on the importance of causality in social interaction, i.e. focusing on human action and interaction, not isolated mental processes. "The main concern is not how the human mind creates the meaning of symbols, but how they learn it during interactions in general and during the socialization process in particular" [21]. One aspect that is put forward in discussing social interaction is socialization. For symbolic interactionists, "socialization is a dynamic process that enables people to develop the ability to think, to grow humanly [22]. This is what Christopher [23] calls the interaction of this dynamic symbol with a "dance" involving a partner. For Ritzer and Goodman [24] interaction is a process when thinking skills are developed and expressed. All kinds of interactions, not just interactions during socialization, polish our thinking abilities. Beyond that, thinking builds the process of interaction.

The use of symbols allows actors to transcend time, space, and even their own person. Through the use of symbols, actors can imagine what it would be like to live in the past or what it would be like to live in the future. In addition, actors can go beyond themselves symbolically and imagine what the world would be like from someone else's point of view. This is the most well-known symbolic interactionist concept, namely taking the role of another [25]. In the process of symbol interaction, it is the person who communicates the meaning symbolically to the other people involved. Others interpret the symbols and direct-action responses based on their interpretation. In other words, in the interaction of symbols, actors are involved in influencing [23]. For symbolic interactionists, actors have at least some autonomy. They are not simply restrained or directed; they are able to make unique and independent choices. In addition, they are able to develop a life that has a unique style (Perinbanayagam, in Ritzer and Goodman, [23]).

2. The meaning of social change

Social change according to Gillin and Gillin in Soekanto [26] is "a variation of accepted ways of life, either due to changes in geographical conditions, material culture, population composition, ideology or because of diffusion or new discoveries. in society". In line with this understanding, Koening argues that "social change refers to the modifications that occur in the patterns of human life that occur due to internal and external causes".

Sztompka [27] in his book Sociology of Social Change, details the definition of socio-cultural change as follows:

1. Social change is a transformation in the organization of society, in the mindset, and in behavior at a certain time.

2. Social change is a modification or transformation in the organization of society.

3. Social change refers to the variation of relationships between individuals, groups, organizations, cultures, and societies at a particular time.

4. Social change is a change in behavior patterns, social relations, institutions, and social structures at a certain time.

Soekanto [28] sees the impact of technological and economic conditions as the cause of social change. This can be seen from his statement that "these conditions are considered the basis of social organization and values are the result of the impact of technological and economic situations". Meanwhile, Astrid sees "change in society in a broad sense defined as change or development in a positive or negative sense. The meaning of social change has two dimensions, namely social change as regression and as progress [29].

3. Coffee shop in Indonesia

The proliferation of coffee shops in big cities in Indonesia cannot be separated from the currents of modernization and globalization of the world. Modernization is characterized by the emergence of new consumption tools, mostly United States innovations that have not only transformed consumption in the United States but are also exported aggressively to most other parts of the world where they have an even greater impact on consumption [30]. Modernization tends to expand its network of reach, especially its space, and this is what is called globalization. For Giddens in Martono, [31] modernity is:

> *Globalization, means that it tends to cover an increasingly wider geographical area and eventually covers the entire world. Modernity also reaches individual personal aspects (beliefs, religion, behavior, consumption tastes, entertainment patterns, and others).*

Dynamic urban communities with all forms of busyness really need the means to relax and release fatigue, emotions, and stress in the world of work. This is what Sztompka alluded to in Martono [32] that:

> *The hallmark of modern society is the separation of time for work and time for leisure, but more time for leisure. Modern human activities are very dense, forcing them to spend a relatively long time just to rest.*

The presence of a coffee shop that provides a relaxed and relaxed atmosphere is an alternative choice for city residents to relax. The development of coffee shops is no longer only caused by the culture of consumerism that develops in modern society but is also caused by a shift in the distribution of social prestige. As stated by Martono [33] that:

> *Social prestige or prestige can be manifested in various ways in modern society. Social prestige is not only manifested in various physical symbols, such as how to dress, or through various attributes attached to a person but the designation of individual status can be seen from non-physical symbols, for example, the choice of places to eat, shopping places, recreation areas, clothing brands used. worn, the language used to communicate, and mastery of technology can indicate a person's status.*

The development of coffee shops in Indonesia in the last decade cannot be separated from the influence of the mass media. The increase in number and its even distribution in almost all major cities in Indonesia is one of the justifications that the mass media, especially television, has turned the world into a global village. Instant and worldwide communication give substance to Marshal McLuhan's idea which was first expressed in the 1980s that the world will become a global village. So it is only natural that Hanners [34] argues that "Western culture will dominate throughout the world. The whole world will become a copy of the lifestyles, consumption patterns, values and norms and ideas and beliefs of Western society".

4. Coffee shops and lifestyle

The phenomenon of the rise of coffee shops in various corners of cities in Indonesia (including coffee shops and cafes that provide coffee) both international coffee shops and local coffee shops, can be said to be the influence of opening an international coffee shop franchise with the largest network in the world, Starbucks, around 1998 with 11 branches in major cities in Indonesia. This has had a major impact on the lifestyle of the urban community, especially in terms of enjoying coffee.

The development of coffee shops in Makassar city, both in terms of quantity and quality, has opened up space for the people of Makassar city to socialize. The patterns of interaction and communication created in coffee shops are so relaxed, loose, and fluid, that cultural boundaries seem to be melting. Various forms of negotiation, conflict resolution, business transactions, as well as consolidation, and political imagery can even be completed in the coffee shop space. This phenomenon makes today's coffee shops a new asset for the people of Makassar city which deserves appreciation for being able to become a bridge for the development of local wisdom, namely tudang sipulung.

When viewed from its cultural-historical roots, coffee shops are not impossible to appear as a substitute for entertainment, a place to hang out, and a place to share information for the people of Makassar city. Therefore, the cultural process that develops so naturally must be appreciated because it has an important role in the transformation of space which is currently being symptomatic in almost all corners of Makassar city.

The existence of a coffee shop in Makassar city has created a new identity, through the creation of a space where Makassar city residents from various economic, educational, and social status backgrounds meet. Coffee shops are also a means of glue for the cultural diversity of the people of Makassar city. But on the other hand, the creation of community centers of this kind is also part of a process of cultural escalation towards the search for a new identity, especially for the younger generation. Chaney [35] in his book Lifestyle argues that "a lifestyle or ways of behaving that are related to conventional expectations which then form new patterns of choice through ways of taste patterns that shape and support the hierarchy of privileges and status".

The culture of enjoying coffee in coffee shops can be categorized as a popular culture when referring to Agger's opinion, according to him, popular culture has a lot to do with everyday problems that can be enjoyed by all people or certain people [36]. Popular culture has more influence on groups of young people and is the center of society's ideology and culture, even though popular culture continues to be a contradiction and debate [36]. Popular culture also becomes part of elite culture in certain societies and shows more of the entertainment side, which then seems more consumptive. This is what underlies Dyer by arguing that "entertainment is a personal need of the society that has been influenced by the capitalist structure" [37].

Popular culture can be grouped into several streams, namely: (a) culture is built based on fun but not substantial, and relieves people from the boredom of working all day long, (b) popular culture destroys traditional cultural values, (c) culture becomes a big problem in the view of Marx's capitalist economy, and (d) popular culture is a culture that drips from above [38]. The culture of drinking coffee for urban people with modern characteristics is closely related to the view of time. Sztompka [39] sees "modern consumer society is marked by shopping and holiday seasons. As a result, the date, apart from having a calendar meaning, is also culturally meaningful for certain people."

Sztompka [40] describes the characteristics of modernity in everyday life which can be seen in the following phenomena:

> *(3) Separation between work and leisure time, and more leisure time, (4) increasing consumerism. Daily life is focused on income and consumption of goods that are considered as symbols of an important role (conspicuous consumption, shopping as a self-satisfying activity regardless of the real need to buy).*

The culture of consumerism in modern society can be seen from several aspects. Consumerism is a form of expansion of capitalist commodity production which has led to an increase in the widespread accumulation of material culture in the form of consumer goods and places of purchase and consumption. This has led to the growth of consumption activities and the prominence of the use of leisure time in contemporary western society.

Meanwhile, the sociological perspective sees that a person's satisfaction obtained from consumed goods is related to his socially structured access. The focus of this perspective lies in the various ways people use goods, in order to create social bonds or social differences. Another perspective is the emotional enjoyment of consumption activities and especially the places of consumption activities that cause arousal and aesthetic pleasure directly to the body. What unites a number of writers are the attention to consumption as a communicative activity rather than an instrumental activity. The focus of attention on images, signs, and symbols of consumption also causes renewed interest in personal identity rather than collective practice [41].

On the other hand, Bourdieu in Scott [41], sees consumption as motivated by the need for social groups to achieve status through differences that strengthen the class position. For Bourdieu, "a sense of judgment is rooted in habitus, is a marker of social class and is closely linked to the hierarchy of access to economic capital, cultural capital and social capital". This is what underlies Scott's statement that consumerism emphasizes consumer culture and personal identity.

4.1 Coffee shop as public space for Makassar city community

Discussions and debates in public life show that democracy needs space to share experiences and correct its implementation in state life. This is what Jurgen Habermas calls the public sphere. The public sphere for Kellner is a place for information, discussion, contestation, political struggle, and organization, which includes broadcast media and new cyberspace, as well as face-to-face interactions in everyday life. What is meant by public space here is all organs of information and political debate, such as newspapers and journals, as well as political discussion institutions, such as parliaments, political clubs, salons and public assemblies, drinking places and coffee shops, meeting halls, and public spaces. Another public space, where discussions on social, political issues can take place.

Habermas identifies the public sphere as an area of social life that ensures access for all citizens to interact, conduct free rational discussions, identify common problems and, through these discussions, influence political action. In Habermas' view, the public sphere is:

> *An arena that is free from the government (even if it is financed by the government) and also one that enjoys autonomy from partisan economic forces, devoted to rational debate (i.e. for debate and discussion that is not based on interests, disguises, and manipulation), and accessible and supervised by the public. It is here, in this public space, that public opinion is built (Imron Rosidin).*

Habermas explained that the "public sphere refers to the functions of criticism and control of state authorities which are carried out by the public informally, as well as formally during periodic elections" (Imron Rosidin).

Habermas argues that in seventeenth and eighteenth century Europe (and, in particular, in England), the gradual spread of capitalism made possible the emergence of a different kind of public sphere: the "bourgeois public sphere". The bourgeois public sphere is increasingly becoming a significant part of social life, made up of private individuals gathered together in public places (such as the coffee shops of the seventeenth century in London, the salons of France, and the table societies of Germany. in the eighteenth century) to elaborate on the key issues of the time (particularly political affairs) and exchange views and opinions on matters important to the common good.

The position of public space for urban communities is very important, considering its function in providing space for people to interact and socialize with social and cultural issues freely without pressure. The development of coffee shops in Makassar city cannot be separated from the influence of modernization and globalization which is strengthened by the mass media. The coffee shop that is becoming a trend in the United States has exported its image to Indonesia. Modern coffee shops that have sprung up and dominate the current coffee shop business in Makassar city are stalls managed by owners of large capital. With adequate capital, good management, such coffee shops are able to exist in the midst of coffee shop business competition because they are able to capture market tastes. Meanwhile, the coffee shops that existed earlier, but were not accompanied by adequate capital, management, and management that followed the tastes of the market, then such coffee shops will run in place or even go out of business.

In Makassar city, Warkop Phoenam (“*Warkop” in Indonesian Term as Warung Kopi, in English: Coffee Shop)* is one of the most phenomenal warkop. Others call it Warkop Dottoro, Dg. Sija, Dg. Anas, Cappo, Warkop 76, Lagaligo, Aleta, Ogie, Planet, La Mario, Coffee Zone, Buana, Rally, and others. Meanwhile, the presence of other coffee shops or new coffee shops, by carrying out and offering new concepts and sensations, has also begun to attract interest from Makassar city residents.

5. The socio-cultural entity

From the perspective of a global cultural approach, we see that globalization is driven by cultural uniformity based on the mass media so that it threatens culture and national identity. This is what underlies McLuhan's term global village with the basic idea being that "the spread of mass media, especially television and now the internet, aims that everyone in the world can be exposed, almost instantly, to the same image.

Horton and Hunt [42] see it from the point of view of "whether an action is important or not important".

Values are collective (together) ideas about what is considered good, important, desirable, and considered worthy, as well as about what is considered not good, unimportant, unwanted, and inappropriate in a culture. Values refer to things that are important in human life, both as individuals and as members of society. There are three classifications of social values: (1) material value, namely, everything that is useful for the body or tangible objects that can be used as human physical needs, (2) vital value, is everything that is useful for humans so that they can carry out activities or activities in his life, (3) spiritual values, (spiritual) humans that are universal.

The coffee shop that exists in Makassar city today is a coffee shop that has experienced a shift in meaning from a traditional coffee shop that only offers coffee as a means of mere physical need to a modern coffee shop that offers sensation and atmosphere (relaxed atmosphere, sexy waiters, internet facilities). Free and various other supporting facilities). So that visitors are happy to spend their time at the coffee shop. In modern coffee shops, it is easy to find visitors who carry out socialization or political negotiations, business transactions, free and fluid discussions of social and cultural issues, as well as those who use coffee shops as a means of releasing congestion and fatigue caused by pressure in the world of work.

The presence of a coffee shop as a means of tudang sipulung for the residents of Makassar city has had an impact on the young people of the city. The coffee shop, which was previously characterized by the majority of visitors from the elderly, has now turned to the young world. Young people as the center of an ideological change, with the presence of coffee shops, slowly but surely have an effect on the increase in consumerism among them. However, the presence of a warung should be seen as something that is beneficial for the social life of the diverse people of Makassar city. Coffee shops should be able to become a means of socio-cultural glue for city residents and a good educational tool for young people.

5.1 Changes in the meaning of coffee shops as space for socio-cultural entities

The rapidly increasing population of Makassar city makes the need for public spaces also increase (**Figure 1**). This potential seems to be well captured by entrepreneurs by opening a coffee shop. The development of coffee shops in Makassar city is more or less related to the culture of the people of Makassar city. H. Haerullah stated:

> *The development of coffee shops in Makassar City cannot be separated from the culture of the Bugis-Makassar people who really like to hang out even though they seem to tend to waste time (interview, January 30, 2017).*

Nawi BS, an IT practitioner who sometimes hangs out at a coffee shop 7 days a week, sees:

> *The emergence of coffee shops in Makassar City is because South Sulawesi people like to get together and like to drink coffee (interview 17 January 2017).*

A coffee shop for coffee connoisseurs is a new public space and social space that allows its visitors to socialize with fellow visitors, both from the community and outside the community. From the habit of visiting coffee shops, coffee shop visitors can make many friends or acquaintances. The intensity of visits to coffee shops is high, making them individuals who are easy to accept new friends. The coffee shop is a unifier in the midst of the diversity of Makassar city residents. For Muhammad Yusuf, this function should receive attention and continue to be developed, as in his following statement:

> *The social benefits of a coffee shop are as a unifier, socializing in a coffee shop without looking at SARA elements (interview on January 15, 2017).*

The improvement of services and facilities provided by coffee shop managers in Makassar city today emphasizes the function of coffee shops as a socio-cultural space for the people of Makassar city. The provision of space with adequate air conditioning facilities makes coffee shops a viable choice for various elements of society. Not infrequently we meet local authorities holding meetings to socialize their programs to both their subordinates and the community or coffee shop community.

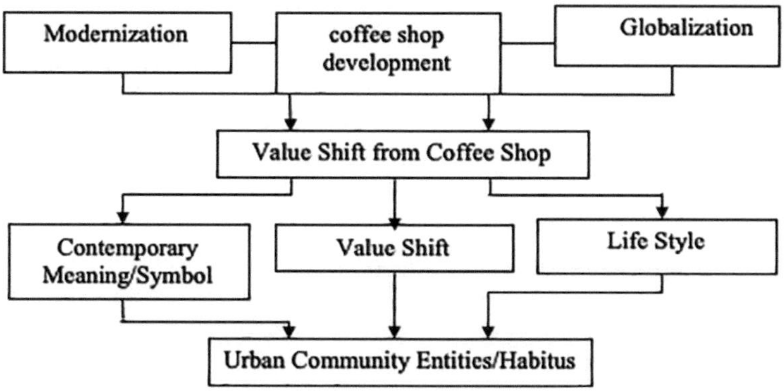

Figure 1.
The dimensions of change in socio-cultural entities.

In addition to discussions in formal forms that are usually carried out by local authorities and company leaders to their employees. In fact, the discussions that are much livelier are those carried out by various communities at their respective tables. The themes raised as topics of discussion are generally situational, depending on events or social realities currently being hotly discussed by the community at that time.

The phenomenon that is also currently developing towards the use of coffee shop space as a social space is bazaar activities by students, birthdays, and reunions. Ahmad Syamsuddin stated that:

> *In addition to interacting as ordinary people, I see that coffee shops have been transformed to replace our living rooms. In the past I wanted to meet an old friend, "let's go home, meet at home", but now, "let's meet at a coffee shop". From some of the writings that I read, the houses that are being built now tend to shrink both in terms of size and in terms of function. In the past, the living room had a social function, interaction with neighbors, family who came, work matters, business. This is almost never done in living rooms but shifts to coffee shop spaces (interview 24 January 2017).*

From the observations, it was found data that it is not uncommon for disputes that arise as a result of social or business relationships to be resolved in coffee shops. Which problems like this were previously solved at home or in the office.

5.2 Changing the meaning of coffee shops as an effective political means

The diversity of visitors to coffee shops in Makassar city is so obvious. There are all kinds of people in the coffee shop. Coffee shops do not discriminate between ethnicity, religion, and race, do not differentiate between rank, position, and occupation. All mingle in a coffee shop space that is free, relaxed, relaxed, and fluid. This diversity is one of the attractions for politicians to enliven the coffee shop discussions.

This phenomenon has grown rapidly since the direct election of mayors was held in Makassar, and it has been cultivated by most politicians in Makassar city. There are several forms of activities carried out by politicians in coffee shops, including events that are packaged in the form of talk shows which usually take place in collaboration with radio stations, local newspapers, and coffee shop managers. From the writer's observation and involvement in coffee shops during this research, the most common form is the socialization of political actors through photos, posters, and billboards that are often attached to the walls of coffee shops.

Socialization carried out by candidates for rulers or candidates for members of the council, basically got mixed responses for the coffee shop community. For those who are not related to practical politics, the activities carried out by politicians in coffee shops are less effective in influencing the preferences of a voter, as stated by Ahmad Syamsuddin that:

> *Politicians socializing in coffee shops means entering one's spare time. When people are having fun and being happy, they come in and offer an issue or program, making it easier to accept. Personally, it's not very effective at influencing one's preferences (interview 25 January 2017).*

Meanwhile, Muhammad Yusuf saw that the activities of political actors in coffee shops had many benefits. Muhammad Yusuf said that:

> *Personally, at least he treats coffee and cigarettes. At least we know who the candidates will be, we get information about the programs offered. This is a political lesson that is not obtained in college or school (interview on 15 January 2017).*

In the world of politics, coffee shops provide an adequate contribution to education and information about politics for some residents of Makassar city. This happened because the political discussions both formally and informally were so intense in coffee shops. Socialization is usually done by political figures or authorities in coffee shops can be a source of information and reference for the coffee shop community. Likewise, the socialization carried out by the candidates and their volunteers, who will advance in a contest for the seat of power in Makassar city, becomes material or source of information for the coffee shop community about the current condition of Makassar city. This is possible because, in socialization or imaging like this, most of the actors discuss the current condition of Makassar city and ideas for future improvements. The frequent discussions of this kind in coffee shops are also one of the main attractions as news material for journalists to hang out in coffee shops. So it is no exaggeration if Ahmad Syamsuddin, states that:

> *The coffee shop is a place to get news, gather and share information with journalists (interview on January 24, 2017).*

One of the clear differences between the coffee shops of the past and the coffee shops of today is adequate facilities for visitors to access information. The coffee shop managers generally provided accessible information facilities, such as television, newspapers, and internet networks. It is from this facility that coffee shop visitors get information. The information obtained is often a hot topic of discussion in coffee shop spaces.

6. Conclusions

1. The change in the meaning of traditional coffee shops to modern coffee shops is marked by simple facilities that develop into public spaces equipped with facilities and information such as television, newspapers, and internet networks to meet the needs and satisfaction of visitors.

2. Coffee shops have become socio-cultural entities in developing patterns of interaction and interrelation as unifiers in association regardless of race, religion, and so on.

3. Coffee shops have been used as a means of political discussion and self-image, both formally and informally, socializing work programs with political figures or authorities and effectively digging up actual information from visitors.

4. Today's coffee shops have a new entity, namely as a unifier in association regardless of race, ethnicity, religion and as a means of political discussion and self-image, both formally and informally.

Conflict of interest

The authors declare no conflict of interest.

Author details

Maksud Hakim* and Ali Syahban Amir
Sekolah Tinggi Ilmu Ekonomi Yapti Jeneponto (Yapti Jeneponto School of Economics), Jeneponto Regency, South Sulawesi Province, Indonesia

*Address all correspondence to: maksudhakim@gmail.com

References

[1] Rivers WL, Peterson T, Jensen JW. Media Massa & Masyarakat Modern. Terjemahan oleh Haris Munandar & Dudy Priatna. Jakarta: Kencana Prenada Media Group; 2003. p. 39

[2] Ritzer G, Goodman DJ. Teori Sosiologi Modern. Terjemahan oleh Alimandan. Jakarta: Kencana Prenada Media Group; 2011. p. 528

[3] Bahtiar W. Sosiologi Klasik. Bandung: Remaja Rosdakarya; 2006. p. 24

[4] Coffee production in Indonesia [Internet]. 2021. Available from: https://id.wikipedia.org/wiki/Kopi_di_Indonesia [Accessed: 03 March 2021]

[5] Ritzer G, Goodman DJ. Teori Sosiologi Modern. Terjemahan oleh Alimandan. Jakarta: Kencana Prenada Media Group; 2011. p. 522

[6] Ritzer G, Goodman DJ. Teori Sosiologi Modern. Terjemahan oleh Alimandan. Jakarta: Kencana Prenada Media Group; 2011. p. 521

[7] Ritzer G, Goodman DJ. Teori Sosiologi Modern. Terjemahan oleh Alimandan. Jakarta: Kencana Prenada Media Group; 2011. p. 523

[8] Ritzer G, Goodman DJ. Teori Sosiologi Modern. Terjemahan oleh Alimandan. Jakarta: Kencana Prenada Media Group; 2011. p. 524

[9] Ritzer G, Goodman DJ. Teori Sosiologi Modern. Terjemahan oleh Alimandan. Jakarta: Kencana Prenada Media Group; 2011. p. 527

[10] Ritzer G, Goodman DJ. Teori Sosiologi Modern. Terjemahan oleh Alimandan. Jakarta: Kencana Prenada Media Group; 2011. p. 529

[11] Martono N. Sosiologi Perubahan Sosial (Perspektif Klasik, Modern, Posmodern, dan Poskolonial). Jakarta: Rajawali Pers; 2011. p. 30

[12] Bahtiar W. Sosiologi Klasik. Bandung: Remaja Rosdakarya; 2006. p. 243

[13] Soekanto S. Sosiologi Suatu Pengantar. Jakarta: Rajawali Pers; 2010. p. 54

[14] Soekanto S. Sosiologi Suatu Pengantar. Jakarta: Rajawali Pers; 2010. p. 58

[15] Bahtiar W. Sosiologi Klasik. Bandung: Remaja Rosdakarya; 2006. p. 248

[16] Soekanto S. Sosiologi Suatu Pengantar. Jakarta: Rajawali Pers; 2010. p. 65

[17] Faisal A. Ruang publik phoenam sebagai budaya politik kontemporer makassar (suatu pertarungan ideologis menuju hegemoni) [Tesis]. Jakarta: Universitas Indonesia; 2008

[18] Rivers WL, Peterson T, Jensen JW. Media Massa & Masyarakat Modern. Terjemahan oleh Haris Munandar & Dudy Priatna. Jakarta: Kencana Prenada Media Group; 2003. p. 28

[19] Bahtiar W. Sosiologi Klasik. Bandung: Remaja Rosdakarya; 2006. p. 247

[20] Bahtiar W. Sosiologi Klasik. Bandung: Remaja Rosdakarya; 2006. p. 249

[21] Ritzer G, Goodman DJ. Teori Sosiologi Modern. Terjemahan oleh Alimandan. Jakarta: Kencana Prenada Media Group; 2011. p. 291

[22] Ritzer G, Goodman DJ. Teori Sosiologi Modern. Terjemahan oleh Alimandan. Jakarta: Kencana Prenada Media Group; 2011. p. 393

[23] Ritzer G, Goodman DJ. Teori Sosiologi Modern. Terjemahan oleh Alimandan. Jakarta: Kencana Prenada Media Group; 2011. p. 396

[24] Ritzer G, Goodman DJ. Teori Sosiologi Modern. Terjemahan oleh Alimandan. Jakarta: Kencana Prenada Media Group; 2011. p. 394

[25] Ritzer G, Goodman DJ. Teori Sosiologi Modern. Terjemahan oleh Alimandan. Jakarta: Kencana Prenada Media Group; 2011. p. 395

[26] Soekanto S. Sosiologi Suatu Pengantar. Jakarta: Rajawali Pers; 2010. p. 263

[27] Sztompka P. Sosiologi Perubahan Sosial. Jakarta: Kencana Prenada Media Group; 2005. p. 5

[28] Soekanto S. Teori Sosial tentang Perubahan Sosial. Jakarta: Ghalia Indonesia; 1984. p. 30

[29] Garna YK. Teori-teori Perubahan Sosial. Bandung: PPS Universitas Padjajaran; 1992

[30] Ritzer G, Goodman DJ. Teori Sosiologi Moderna. Terjemahan oleh Alimandan. Jakarta: Kencana Prenada Media Group; 2011. p. 570

[31] Martono N. Sosiologi Perubahan Sosial (Perspektif Klasik, Modern, Posmodern, dan Poskolonial). Jakarta: Rajawali Pers; 2011. p. 82

[32] Martono N. Sosiologi Perubahan Sosial (Perspektif Klasik, Modern, Posmodern, dan Poskolonial). Jakarta: Rajawali Pers; 2011. p. 84

[33] Martono N. Sosiologi Perubahan Sosial (Perspektif Klasik, Modern, Posmodern, dan Poskolonial). Jakarta: Rajawali Pers; 2011. p. 89

[34] Hakim M. Nelayan dan Eksploitasi Struktur. Oase Pustaka: Surakarta; 2016. p. 55

[35] Chaney D. Life Styles. Sebuah Pengantar Komprehensif. Terjemahan Oleh Nurnaeni. Yogyakarta: Jalasutra; 2003. p. 35

[36] Bungin B. Konstruksi Sosial Media Massa. Jakarta: Kencana; 2011. p. 49

[37] Bungin B. Konstruksi Sosial Media Massa. Jakarta: Kencana; 2011. p. 50

[38] Bungin B. Konstruksi Sosial Media Massa. Jakarta: Kencana; 2011. p. 4

[39] Sztompka P. Sosiologi Perubahan Sosial. Jakarta: Kencana Prenada Media Group; 2005. p. 63

[40] Sztompka P. Sosiologi Perubahan Sosial. Jakarta: Kencana Prenada Media Group; 2005. p. 89

[41] Scott J. Sosiologi (The Key Concepts). Terjemahan oleh Tim Penerjemah Labsos FISIP UNSOED. Jakarta: Rajagrafindo Persada; 2011. p. 62

[42] Horton & Hunt. Sosiologi. Jakarta: Erlangga; 1979. p. 95

Chapter 2

High-Intensity Ultrasound and Its Interaction with Foodstuff and Nanomaterials

Ana Luisa Rentería-Monterrubio, América Chávez-Martínez, Julianna Juárez-Moya, Rogelio Sánchez Vega, Juan Manuel Tirado and Raúl Alberto Reyes-Villagrana

Abstract

In recent decades, food research has focused on hybrid systems, that is, the application of nanomaterials and the so-called emerging technologies, whose objective is to increase the quality of food. Among these technologies that are different from thermal is high intensity ultrasound. This chapter presents and describes the interaction of acoustic waves supported by classical physics and nanomaterials generated by nanotechnology carried out in contemporary physics, all integrated as a multidisciplinary knowledge applied to food. Acoustic waves have a spectrum called ultrasound, with an approximate frequency range between 20 kHz and 1 GHz, and this is divided into low-intensity ultrasound (LIU), whose objective is to characterize materials. On the contrary, high-intensity ultrasound (HIU) has the objective of carrying out destructive tests, modifying the study samples. This happens because the HIU generates a phenomenon called acoustic cavitation, which consists of the generation, growth, and implosion of microbubbles, causing alterations in the near and far acoustic field. The proposed review focuses on the application of high-intensity ultrasound to be used in the food industry. Subsequently, a brief approach is made to nanotechnology and nanomaterials and how they have been incorporated into the food industry.

Keywords: acoustic waves, acoustic cavitation, dairy, foods, high-intensity ultrasound, meat, nanotechnology, nanomaterials, ultrasound

1. Introduction

With the evolution of the human being from the time of the Homo's, they tried to survive and explore nature (the environment). In 1859 Charles Darwin (1809–1882) published his book entitled the origin of the species [1], where he described that the species that would survive be those that adapted more quickly to its environment or the change of the environment. In such a way that a struggle for survival was generated and later for the observation of nature, learning, and understanding of it, answering questions that were raised and that nowadays, the scientific community continues to make new questions about it.

The advance of science and technology has grown thanks to a set of multidisciplinary groups generating knowledge and developing what we know as technology, however, some scientists have excelled for their most relevant contributions. This is the case of the area of physics as a cornerstone in the strengthening of science, which has had transcendental epochs, so to speak, since the time of ancient, classical, modern, and contemporary physics. For example, in ancient Greece, the knowledge developed and established up to that time was manifested by the natural philosophy of Aristotle (384–322 AC). At that time Democritus (460–370 BC) argued for the first time the word atom, which means indivisible and which he described as the smallest part that made up all "things." This was the first approach to the description of very small things. Later in classical physics, the observations of Galileo Galilei (1564–1642) and Isaac Newton (1643–1727) laid the foundation of the behavior of bodies (dynamics). It was the latter who with his book entitled Mathematical Principles of Natural Philosophy [2], established the behavior of different natural phenomena under mathematical arguments. Similarly, the observations made in the laboratory by Michael Faraday (1791–1867) and the dazzling capacity of James Clerk Maxwell (1831–1879), unify the electricity and magnetism manifested in treaties bearing the same name [3]. And so-the transformation of natural philosophy to physics, classical physics, was presented.

After this, new contributions to knowledge emerged, as did J.J. Thomson (1886–1943), H. Becquerel (1852–1908), M. Plank (1858–1947), E. Rutherford (1871–1937), N. Bohr (1885–1962) E. Schrödinger (1887–1961), J. Chadwick (1891–1974), E. Fermi (1901–1954), and A. Einstein (1879–1955) and the entire generation of modern physics, where they laid the foundations of quantum mechanics and special and general relativity, that is, small particles moving at speeds of light and their interaction with gravity. And in that time a leap was generated in the era of electronics and the advancement of technology with the invention of the transistor made by J. Bardeen (1908–1991), W. H. Brattain (1902–1987), and W.B. Shockley (1910–1989) in 1947. Toward the macroscopic nature, and then microscopic, and later quantum, new questions and advances given by the contemporary physics were presented by P. Dirac (1902–1984), H. Yukawa (1907–1981), J.S. Schwinger (1918–1994), R.P. Feynman (1918–1988), M. Gell-Mann (1929–2019), S. Weinberg (1933–2021), P. Higss (1929-), the world of elementary particles and their interactions. To generate, this knowledge also originated the technological development implemented in the European Center for Nuclear Research leading to a modern technological development [4].

The need to address gaps in knowledge for the development and manipulation of "very small" things led to the beginning of nanotechnology and its possible applications for the benefit of humanity, as it is in medicine, electronics, food, etc. This chapter presents the interactions of classical physics represented by acoustic waves and modern physics implemented in nanotechnology. All this is integrated as a multidisciplinary knowledge applied to food. This chapter describes the basic notions of acoustic waves and consequently of low and high-intensity ultrasound to continue with the effect of acoustic cavitation. Later, the progress of nanotechnology and how it has been incorporated into the food industry is addressed.

2. Acoustic waves

Acoustic waves have a wide field of application and can be interpreted as the physical science that studies the vibrations and elastic waves, and their interactions with the environment; both in macroscopic systems and in quantum systems (phonons). At the microscopic level, acoustic waves have an intermolecular collision process and unlike electromagnetic waves, a material medium is needed for the propagation of the acoustic wave. At the macroscopic level, acoustic waves have an intermolecular collision process and unlike electromagnetic waves, a material medium is needed for the propagation of the acoustic wave. At the macroscopic level, acoustic waves refer to variations in the amplitude of pressure-dependent on time or stress, so acoustic waves are also known as mechanical waves or elastic waves [5].

The linear wave equation is represented by the expression,

$$\nabla^2 P(\boldsymbol{r},t) - \frac{1}{c^2}\frac{\partial^2 P(\boldsymbol{r},t)}{\partial t^2} = 0, \quad (1)$$

where ∇^2 is the three-dimensional Laplacian operator, which can be applied to different geometries, and this will depend on the type of wave that is analyzed, which can be flat, cylindrical, and/or spherical waves. P is the acoustic pressure of the wave. $\boldsymbol{r}$ is the position vector of the wave and is given by $\boldsymbol{r} = x\hat{i} + y\hat{j} + x\hat{k}$, in cartesian coordinates. $\hat{i}, \hat{j}, \hat{k}$, they are the unit vectors. c is the longitudinal velocity of the propagation medium, t is time, ∂^2 represents the second partial derivate of the argument. The expression (1) represents a homogeneous wave equation, that is, it does not matter what or who generates the acoustic wave.

The solution of the harmonic flat wave equation is represented by the expression,

$$P = \boldsymbol{A}e^{-i(\boldsymbol{k}\bullet\boldsymbol{r}-\omega t)} + \boldsymbol{B}e^{i(\boldsymbol{k}\bullet\boldsymbol{r}-\omega t)}, \quad (2)$$

where $\boldsymbol{A}$ and $\boldsymbol{B}$ are constants that can be determined by applying initial and/or boundary conditions; k is the wave propagation vector and is given by $\boldsymbol{k} = k_x\hat{i} + k_y\hat{j} + k_z\hat{k}$, i is the imaginary number, k is the wavenumber, and it is defined as $k = \frac{\omega}{c}.\omega$ is the angular frequency. Also, c can be represented as $c = \lambda \cdot f$, where λ is the wavelength and f is the frequency. Finally, λ can also be represented by $\lambda = \frac{2\pi}{k}$.

As mentioned in the previous paragraph, acoustic waves need a material medium for propagation, this medium can be solid, liquid, gas, or biological material. There are two main modes for the propagation of acoustic waves, which are longitudinal mode and transverse mode. In the longitudinal mode, the emitted acoustic energy propagates in the same direction in which the acoustic travels; on the contrary, in the transverse mode, the emitted acoustic energy propagates perpendicularly in the direction in which the acoustic wave travels. **Figure 1** describes the main modes of propagation of acoustic waves [6].

Acoustic waves have well-defined parameters in steady state. **Figure 2** shows the parameters of an acoustic wave represented by a sine wave.

In **Table 1** the spectrum of acoustic frequencies is described.

Infrasound is an acoustic wave whose frequency is below the audible spectrum of the human ear; some animals can detect these frequencies. The audible sound is conformed by the spectrum of frequencies that can be perceived by the human ear, such as, for example, the voice, the musical notes. Ultrasound is an acoustic wave

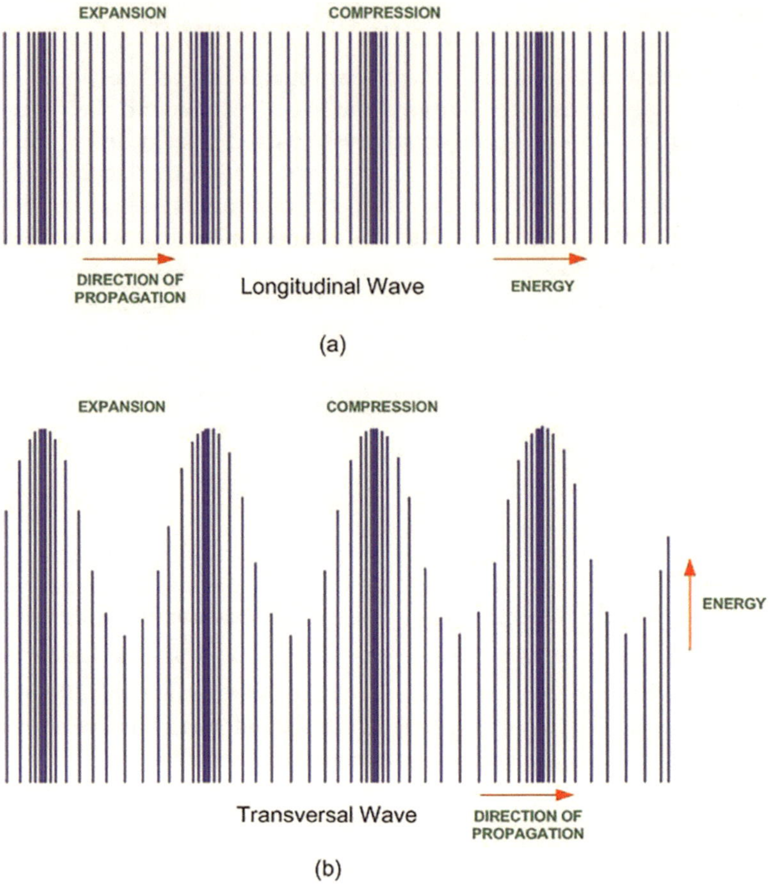

Figure 1.
Propagation modes of acoustic waves, (a) longitudinal mode; (b) transverse mode.

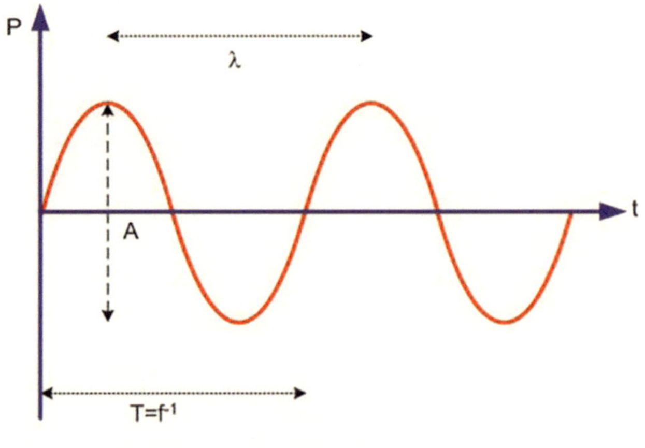

Figure 2.
Sine wave. A→ Amplitude, *λ*→ Wavelength, *T* → Period, *f* → Frequency, *P* → Pressure, *t* → Time. *Where P is the pressure [Pascal], A is the amplitude of the pressure [meters], t is the time [second], λ is the wavelength [meter], T is the period [second] and f is the frequency [Hertz], which* $T=f^{-1}$ *or* $f=T^{-1}$*, and it is the latter that provides the working spectrum of acoustic waves [7].*

Name	Range (Hz)
Infrasound	IS < 20
Audible sound	20 < AS < 20k
Ultrasound	20k < US < 1G
Hypersound	HS > 1G

Table 1.
The spectrum of acoustic frequencies [8].

whose frequency is above the audible spectrum of the human ear; ultrasound has several applications, such as non-destructive and destructive tests in medical diagnosis and rehabilitation therapies. The hypersound is an acoustic wave whose frequency spectrum is very high and has some applications in medicine and war systems.

In this context, acoustic waves also obey Snell's law of optics [9], where the incident, reflected, and transmitted waves are involved in a propagation medium, $\frac{Sin\theta_i}{c_i} = \frac{Sin\theta_t}{c_t}$, where θ_i, θ_t, c_i, and c_t represent the angle of the incident wave, transmitted wave, acoustic longitudinal propagation speed of the incident and transmitted wave, respectively. The phenomenon of attenuation is also presented, which is the loss of energy given the intrinsic properties of the medium of propagation, the frequency, and the thickness thereof, and is represented by the expression,

$$P = \boldsymbol{P_o} e^{-i\alpha r}, \tag{3}$$

where $\boldsymbol{P_o}$ is the amplitude of the pressure, i is the imaginary number, α is the coefficient of attenuation of the medium and are included the losses by absorption, diffraction, dispersion, scattering, as well as the object of study; **r** is the position vector of the acoustic wave, as shown in **Figure 3**.

From **Figure 3**, the following expression can be described,

$$P_i = \tilde{\boldsymbol{P}}_{\boldsymbol{i}} e^{i(\boldsymbol{k}\cdot\boldsymbol{r}-\omega t)}, \tag{4}$$

$$P_r = \tilde{\boldsymbol{P}}_{\boldsymbol{r}} e^{-i(\boldsymbol{k}\cdot\boldsymbol{r}+\omega t)}, \tag{5}$$

$$P_t = \tilde{\boldsymbol{P}}_{\boldsymbol{t}} e^{i(\boldsymbol{k}\cdot\boldsymbol{r}-\omega t)}, \tag{6}$$

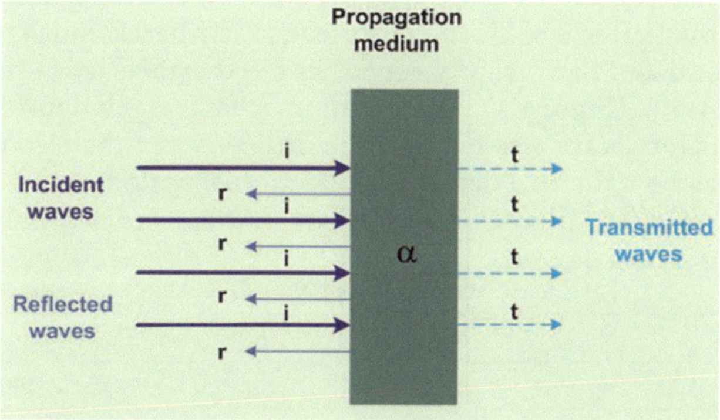

Figure 3.
With normal incidence to the surface of the propagation medium, the acoustic waves hit, are reflected and transmitted. α is the attenuation that exists when an incident crosses the propagation medium.

where P_i, P_r, and P_t, represent the waves of incident waves, reflected, and transmitted. Expressions of pressure with the symbol $\sim$ represent a constant that can be determined by applying initial and/or boundary conditions.

Finally, all the materials have a very important acoustic property, and it is called the specific acoustic impedance (Rayls) that is given by the product of the volumetric density (kg/m^3) and the velocity of propagation (m/s), this is, $Z = \rho \cdot c$. The specific acoustic impedance is the resistance that material presents to the flow of the propagation of the acoustic wave; it should be noted that in solid materials, the longitudinal and transverse acoustic propagation velocities are considered.

2.1 High-intensity ultrasound

Acoustic waves within the ultrasound spectrum range in frequencies between 16kHz and 1GHz. Ultrasound is divided into two categories, according to their acoustic intensities; low-intensity iltrasound (LIU) and high-intensity ultrasound (HIU). LIU is used for exploratory and characterization purposes; meanwhile, HIU is aimed at modifying the biological, physical, and chemical properties of materials. HIU is divided into three subcategories, according to their acoustic intensities depending on their frequency, as described in **Table 2**.

The ultrasound field uses an ultrasonic sensor, which depends on a piezoelectric for its proper operation, ultrasonic sensor has low and high-intensity applications.

2.2 Piezoelectric materials

Piezoelectric materials are those that when subjected to mechanical stress (pressure) generate electrical polarization through the formation of dipoles causing electrical charges of their surface, from which the potential difference can be determined or measured. On the contrary, applying electrical pulses through both surfaces of the material (potential difference), generated mechanical deformations, or the reverse piezoelectricity effect. There are natural piezoelectric materials (quartz), piezopolymers (polyvinylidene difluoride, PVDF), piezoceramics (lead zirconate titanate, PZT), and piezocomposites (PSMNZT) [11, 12]. Piezoelectric materials have three physical properties—mechanical, electrical, and thermal. The relationship between mechanical and electrical properties has an electromechanical effect owning implicitly the piezoelectricity. The association between electrical and thermal properties as electrothermal effects holds the pyroelectric attributes. Finally, the relation between mechanical and thermal properties has the thermoelastic effect and has thermal pressure traits (**Figure 4**). The thermomechanical relationship is remarkable, as piezoelectric materials are also pyroelectric, hence, when the electrical charge changes, they become a thermal detector of electromagnetic waves on the infrared spectrum. However, when piezoelectric property exceeds its thermal threshold, loses

Subcategory	Acoustic intensity range (W/cm^2)	Frequency range (Hz)
Low	$1 \leq L < 2$	$< 1M$
Medium	$2 \leq M < 10$	$100k \leq f < 1M$
High	$10 \leq H < 1000$	$20k \leq f < 100k$

Table 2.
Classification of acoustic intensities in high-intensity ultrasound depending on frequency [10].

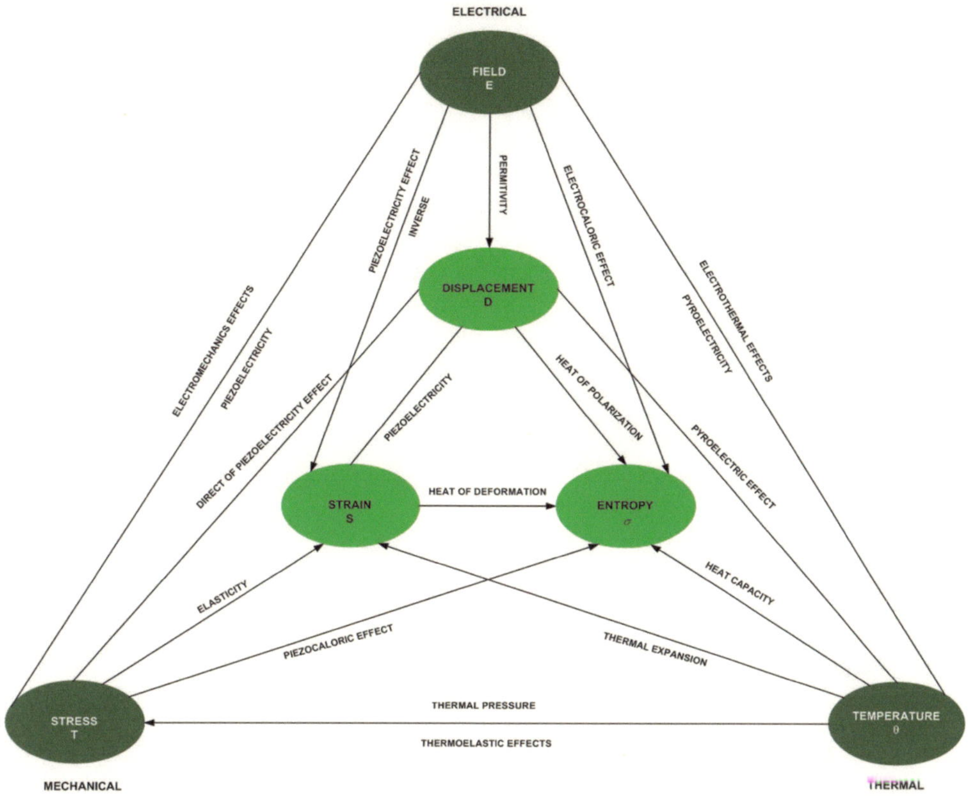

Figure 4.
Heckmann piezoelectricity diagram [13].

its mechanical attributes, and keeps the thermal characteristics, retaining the pyroelectric effect.

Commercial HIU equipment basically has the same inner components. **Figure 5** describes the stages of a HIU system; power supply, signal generator, power amplifier, and the high-intensity acoustic emitter. The high-intensity acoustic emitter has a backing, a pair of ceramic piezoelectric in a sandwich configuration, and an amplifier acoustic horn. The previously described ultrasonic emitters are called Langevin transducers [14].

2.3 High-intensity ultrasound equipment

Ultrasonic baths and ultrasonic emitters (also known as sonotrode or probe types) within academia and research are used for sanitizing purposes (e.g., glassware, spare parts, surgical instruments, ballistics, and among others) because they are easy to handle. There is a wide market for these instruments worldwide, some of these are shown in **Figure 6**.

2.4 Acoustic cavitation

The high-intensity acoustic wave within the ultrasound range frequency, under inertial or transient state conditions, induces a phenomenon known as acoustic

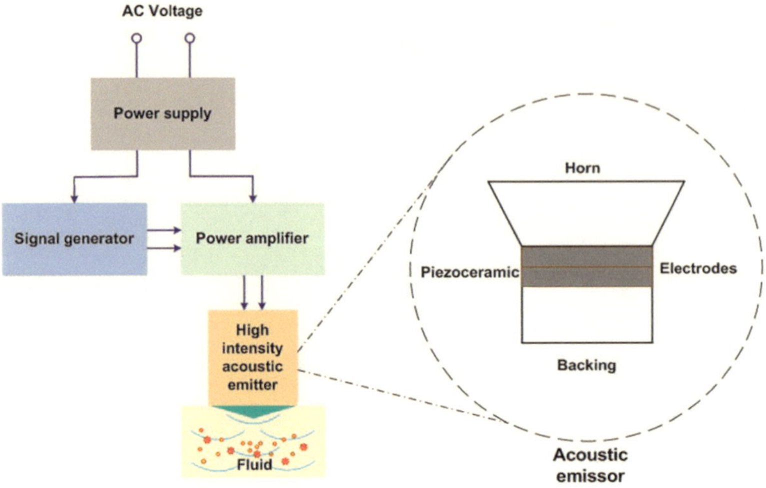

Figure 5.
High-intensity ultrasound stages.

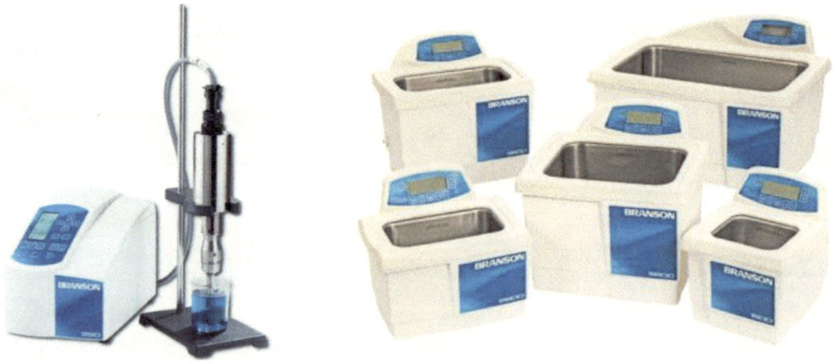

Figure 6.
High-intensity ultrasound equipment [15]. (a) High-intensity ultrasound system, sonotrode type [16]. (b) Ultrasonic bath [17].

cavitation, which includes generation, growth, and collapse of bubbles. The phenomenon occurs in the acoustic wave transition from the negative half cycle to the positive half cycle (expansion and compression). In a stable phase, bubbles generate an increase in number oscillating through the acoustic field (**Figure 7**).

In a transitory state and under stochastics situations, the bubbles collapse causing various effects, such as acoustic microcurrents, nucleation of bubbles, shockwaves, sonoluminescence, radical formation, ultrasonic radiation, streams of cloud- and filamented-shaped bubbles form [18].

There is no general agreement on the area of knowledge that originated the study of the acoustic cavitation phenomenon, however, sonoluminescence, sonophysics, sonochemistry [19–21], and mechanochemistry [22] are the most likely.

The study of acoustic cavitation has developed various physic-mathematical models to describe detailly the phenomenology. The models represent a single bubble under ideal conditions, the description by Rayleigh [23], describes a spherical bubble embedded in an incompressible fluid, where R is the time-dependent radius of the

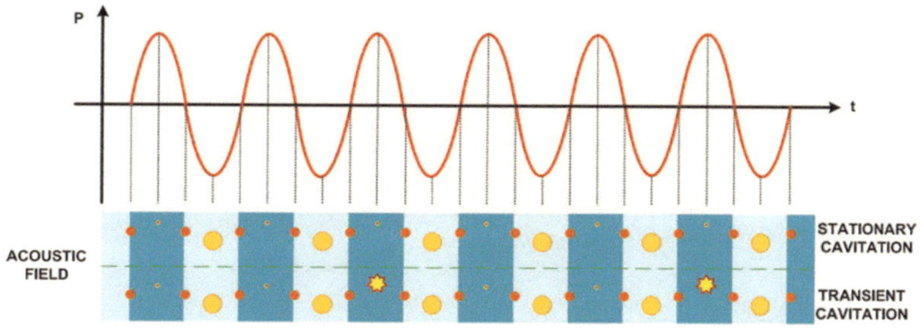

Figure 7.
Description of stable and transient cavitation.

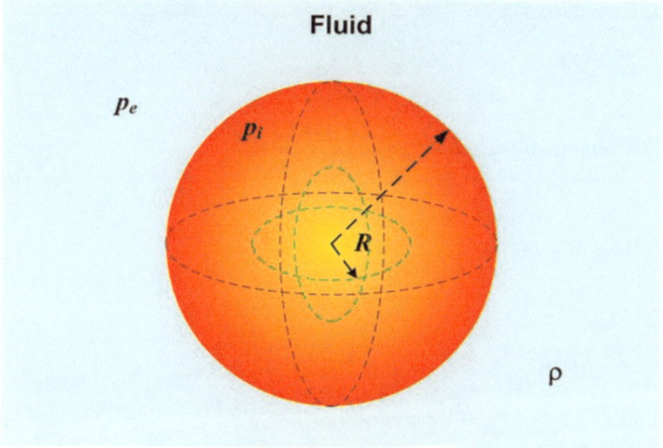

Figure 8.
Physical parameters of the bubble, Rayleigh model.

bubble, $R(t)$, p_e is the external pressure of the fluid. p_i is the pressure inside the bubble, and ρ is the bulk density; see **Figure 8**. This model is represented by Eq. (7) and describes the kinetic energy in the system between the differences in the stress of the bubble as it expands through internal pressure and the stress of the fluid through the pressure subjected to the bubble.

$$\rho R\ddot{R} + \frac{3}{2}\rho\dot{R}^2 = p_i - p_e, \tag{7}$$

where the points on the R (one and two points) mean first and second derivatives with respect to time (Newton's notation), respectively.

The Rayleigh–Plesset model [24] was generated afterward. The parameters κ, μ, σ, from the original Rayleigh model were included, they are the polytropic exponent of the gas inside the bubble, the dynamic viscosity, and the surface tension of the fluid, respectively, as shown in **Figure 9**.

The Rayleigh–Plesset model is represented by Eq. (8),

$$\rho R\ddot{R} + \frac{3}{2}\rho\dot{R}^2 = p_{gn}\left(\frac{R_n}{R}\right)^{3\kappa} + p_v - p_{stat} - \frac{2\sigma}{R} - \frac{4\mu}{R}\dot{R} - p(t), \tag{8}$$

with

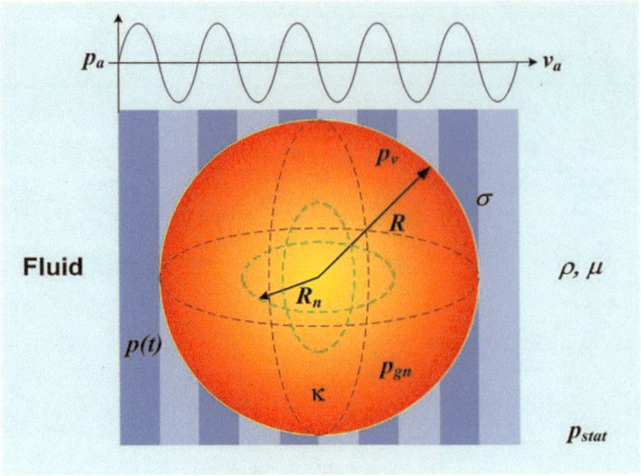

Figure 9.
Physical parameters of the bubble, Rayleigh–Plesset model.

$$p_{gn} = \frac{2\sigma}{R_n} + p_{stat} - p_v, \quad (9)$$

and

$$p(t) = -p_a \sin(2\pi v_a t), \quad (10)$$

where R_n is the radius of the resting sphere. p_{gn} is the pressure of the gas inside the bubble. p_{stat} is the static pressure and p_v is the vapor pressure. $p(t)$ is the external pressure applied to the wall of the bubble. v_a is the frequency and p_a is the amplitude of the pressure.

Later the Gilmore model was developed [25], as represented by **Figure 10** and Eq. (1).

$$\left(1 - \frac{\dot{R}}{C}\right) R\ddot{R} + \frac{3}{2}\left(1 - \frac{\dot{R}}{3C}\right)\dot{R}^2 = \left(1 + \frac{\dot{R}}{C}\right)H + \frac{\dot{R}}{C}\left(1 - \frac{\dot{R}}{C}\right)R\frac{dH}{dR}, \quad (11)$$

with

$$H = \int_{p|_{r\to\infty}}^{p|_{r=R}} \frac{dp(\rho)}{\rho}, \quad (12)$$

$$p(\rho) = A\left(\frac{\rho}{\rho_0}\right)^{n_T} - B, \quad (13)$$

$$p\Big|_{r=R} = \left(p_{stat} + \frac{2\sigma}{R_n}\right)\left(\frac{R_n^3 - bR_n^3}{R^3 - bR_n^3}\right)^{\kappa} - \frac{2\sigma}{R} - \frac{4\mu}{R}\dot{R}, \quad (14)$$

$$p|_{r\to\infty} = p_{stat} + p(t), \quad (15)$$

$$C = \sqrt{c_0^2 + (n_T - 1)H}, \quad (16)$$

DOI: http://dx.doi.org/10.5772/intechopen.104437

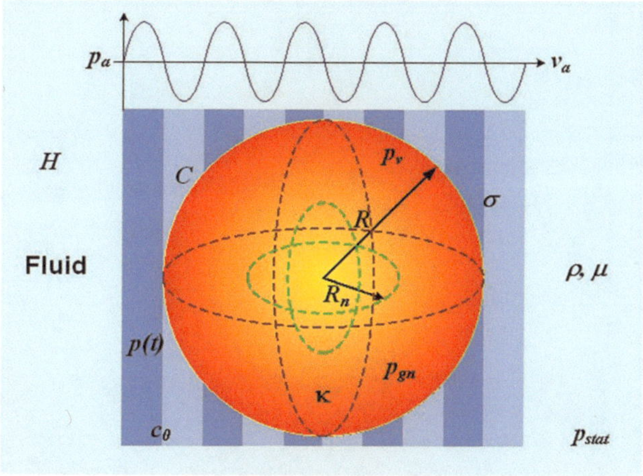

Figure 10.
Physical parameters of the bubble, Gilmore model.

where C is the acoustic velocity near the bubble-wall, c_0 is the acoustic velocity under normal conditions. H is the enthalpy. A, B and nT are the van der Waals parameters.

Finally, the Keller–Miksis model [26] describes the behavior of a bubble under quasi-real conditions, as given by Eq. (17).

$$\left(1-\frac{\dot{R}}{C}\right)R\ddot{R}+\frac{3}{2}\dot{R}^2\left(1-\frac{\dot{R}}{3C}\right)=\left(1+\frac{\dot{R}}{C}\right)\frac{p_l}{\rho}+\frac{R}{\rho c}\frac{dp_l}{dt}, \tag{17}$$

with

$$p_l=\left(p_{stat}+\frac{2\sigma}{R_n}\right)\left(\frac{R_n}{R}\right)3\kappa-p_{stat}-\frac{2\sigma}{R}-\frac{4\mu}{R}\dot{R}-p(t), \tag{18}$$

$$p(t)=p_a\sin\left(2\pi\upsilon_a t\right). \tag{19}$$

Acoustic cavitation in a stable state given the oscillation in the acoustic field generates bubbles and divides them. On the other hand, acoustic cavitation in a transient state refers to the variation of pressure in short periods with respect to time and the breakage of the molecular bonds of the fluid, which will increase rapidly the temperature [27], as seen in **Figure 11**.

If a fluid (liquid) is under shear stress, the bubbles within the fluid increase in size, hence, bubbles are generated within the fluid. The cavities with vapor increase their size until reaching a maximum volume, therefore, when the wave changes pressure from peak to valley, transforms potentials energy into kinetic energy during the implosion. The cavities collapse to sizes even smaller than the originally generated bubbles [28].

The implosions of the bubbles neighboring the surface of an interface are asymmetric; hence, they form microcurrents of the fluid that impinge on the surface of the interface. The average speed of the microcurrents is between 100 and 340 m/s; and they are dependent on the pressure profile and the initial diameter of the bubbles, which ranged between 10 and 100 μm [29].

There are two main types of formations in the generation and accumulation of bubbles in an acoustic field—bubble clouds and bubble filaments, as seen in **Figure 12**.

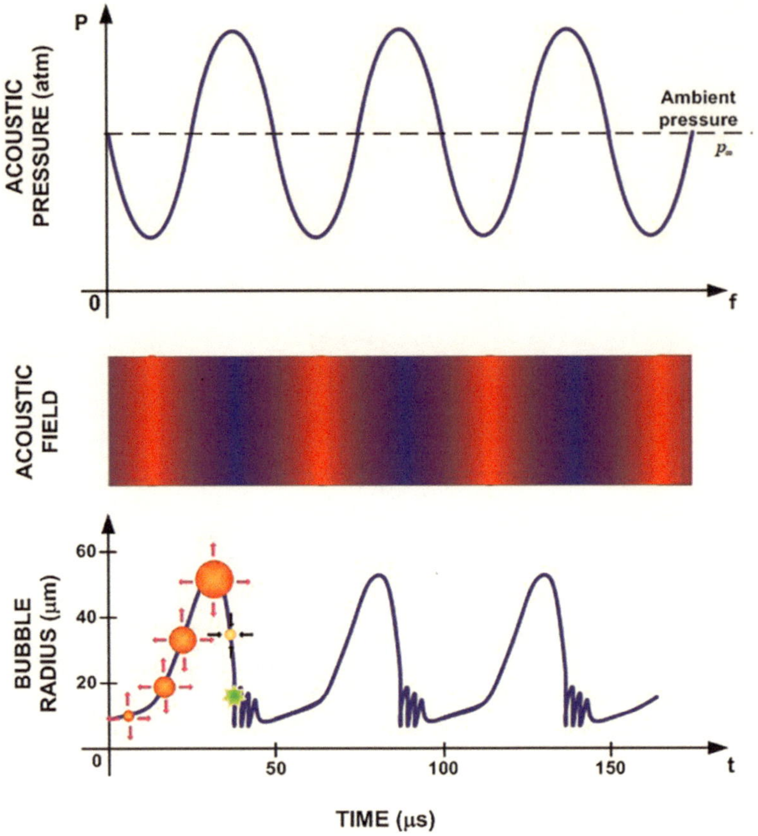

Figure 11.
Bubble generation transitions, growth, and collapse, in a transitory state.

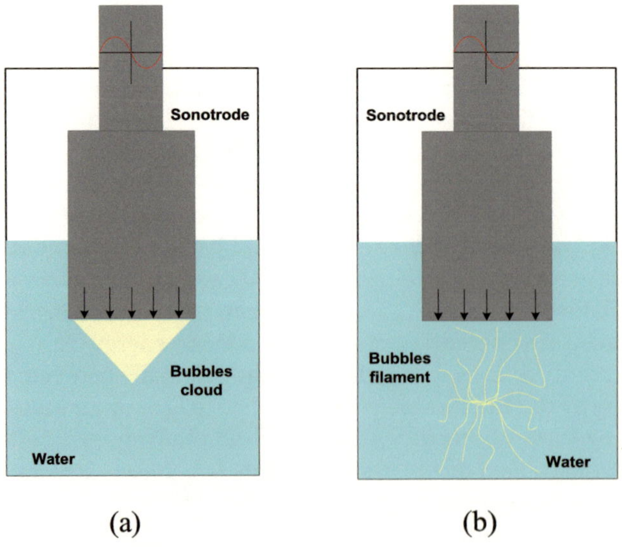

Figure 12.
Multibubble acoustic cavitation. (a) cloud-type, (b) filament-type.

The multibubble acoustic cavitation formation can be represented by the Keller–Miksis model [30], as shown in the Eq. (20).

$$(1-M_i)R_i\ddot{R}_i+\frac{3}{2}\left(1-\frac{M_i}{3}\right)\dot{R}_i^2$$
$$=(1+M_i)\frac{1}{\rho_{li}}\left[p_{li}-p_\infty-p_{si}(t+t_{R_i})\right]+\frac{t_{R_i}}{\rho_{li}}\dot{p}_{li}-\sum_j\frac{\left(2R_j\dot{R}_j^2+R_j^2\ddot{R}_j\right)\Big|_{t-\frac{r_{ij}}{c_{l\infty}}}}{r_{ij}}, \quad (20)$$

where $R_i(t)$ is the radius of the i-th bubble. ρ_{li} is the volumetric density of the fluid outside the i-th bubble. p_∞ is the ambient pressure. $p_{si}(t)=p_{a_i}\sin(\omega t)$, is the modulated acoustic pressure on the i-th bubble. $t_{R_i}\equiv R_i/c_{li}$, cli is the acoustic velocity of the fluid outside the i-th bubble. $p_{li}=p_{gi}(R_i,t)-\frac{4\eta\dot{R}_i}{R_i}-\frac{2\sigma}{R_i}$, is the pressure on the fluid on the side of the i-th bubble wall.

3. Nanomaterials

Metrology states that in the international system of units, the meter [m] is the unit of measure of the length and is defined as the length of the path traveled by the light in the vacuum in a span of 1/299792458 of a second. Its primary standard or reference is at the International Office of Weights and Measures in Sévres, France. In such a way that its submultiples are given by cm, mm, nm, pm, etc. [31].

The unit of measure of the length is the origin to the reference of nanotechnology, due to the dimensions that the latter represents. Nanotechnology can be interpreted with the architecture of functional system and manipulation at nanometric scales, this is one billionth of the meter ($1\times10^{-9}\rightarrow 0.000000001$m) [32]. **Figure 13** shows the comparison of sizes of some objects.

It all started in the American physics congress of 1959 when Richard P. Feynman gave a lecture entitled, "There´s plenty of room at the bottom" [33], and with this, he glimpsed the pioneering work of nanotechnology. Feynman described that it would be possible to manipulate systems at the atomic level individually with high-precision instruments. Also, he would be the first to propose the vision of quantum computing. In this way, the race for the exploration and development of nanotechnology and consequently of nanomaterials began [34–40]. So, a revolution in applications, such as the case of food.

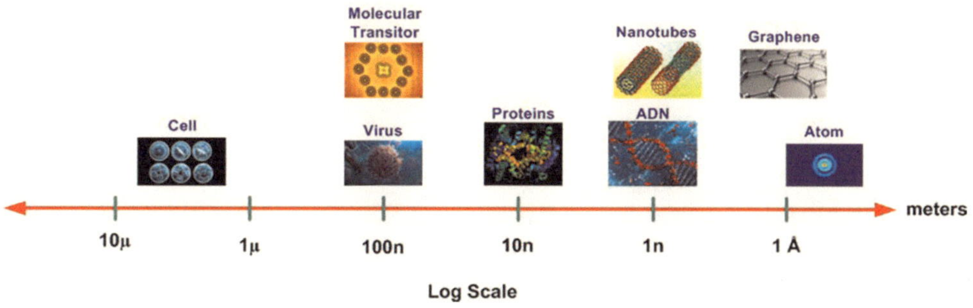

Figure 13.
Dimension comparison.

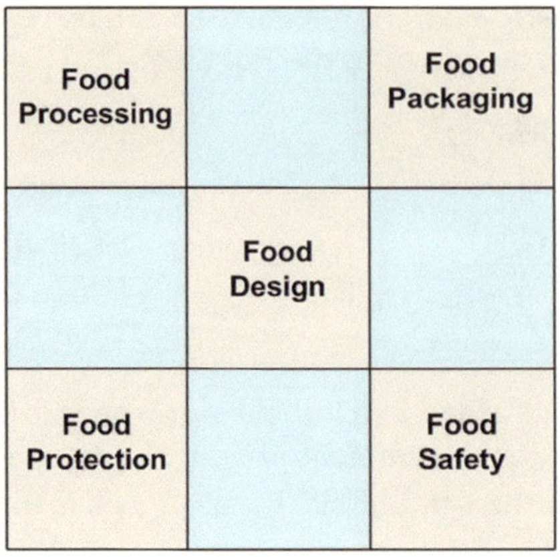

Figure 14.
The five main areas in the study of food.

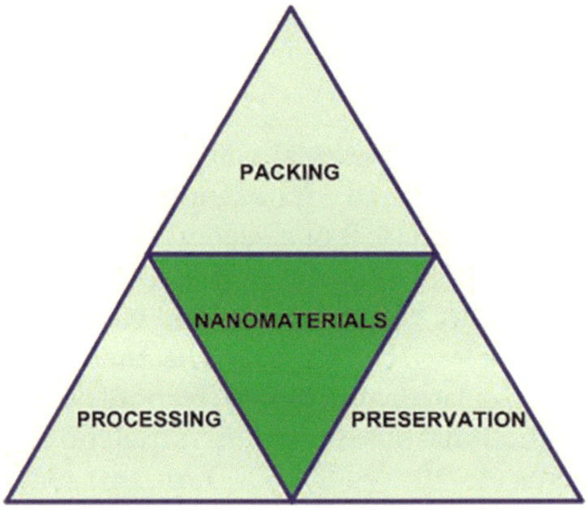

Figure 15.
Applications of nanomaterials in food, by means of the Sierpinski triangle.

There is a wide variety of nanomaterials, however, the main applications are by means of simple nanoparticles, nanoemulsion, liposome, nanostructured lipid carrier, multilayer nanoparticles, agglomerated nanoparticles, nanofilms, and nanocomposites [41]. There are five areas where studies and development in the field of food are focused [42–60], as seen in **Figure 14**.

Subsequently, the interaction of nanomaterials and food interact on three fundamental processes [61–64], as shown in **Figure 15**.

In recent years, high-intensity ultrasound has been applied to foods involving nanomaterials and obtaining interesting results. **Figure 16** shows the interaction of the HIU and the fields of study in nanomaterials focused on food [65–67].

4. Conclusions

Food is a fundamental part of the growth and development of the human being and has changed over time due to current needs. Today the agri-food industry has to use and apply every tool and technology within its reach due to consumer demands. This happens throughout the traceability procedure as it is in the processing, treatment,

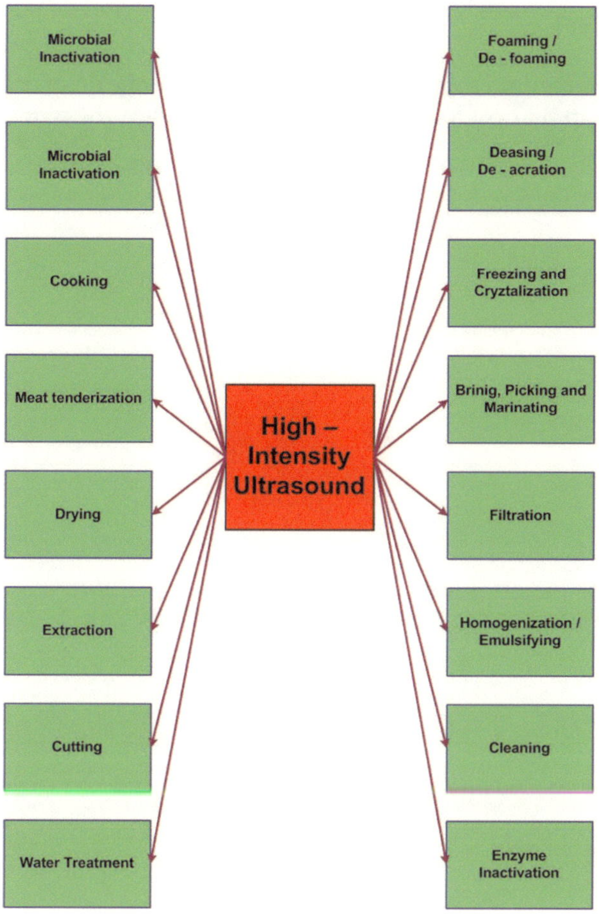

Figure 16.
HIU application in nanomaterials.

conservation, and distribution of food. Therefore, alternative or emerging tools are applied, which are different from thermal ones, whose objective is to increase the nutritional properties of foods with benefit for society. In addition to the conjugation of hybrid systems, such as the implementation of nanomaterials applying modern technologies. In this chapter, high-intensity ultrasound and the effect of acoustic cavitation were explained in a simple way, as a tool that is applied to food. Thus, the application of nanomaterials in the main areas of food was also superficially described. Even so, academic and laboratory studies are still being presented to explore the short-range impacts of these hybrid tools. However, there is the possibility of taking it to large-scale production levels, with the aim of carrying out acceptance and quality tests. Subsequently, the final impacts are to implement it in the food industry worldwide. Also, why not say so, have them as another utensil in every home.

Acknowledgements

R.A.R.V. acknowledges the support provided by Investigador por México-CONACYT.

Conflict of interest

The authors declare that they have no known competing financial interest or personal relationship that could have appeared to influence the work reported in this material.

Author details

Ana Luisa Rentería-Monterrubio[1], América Chávez-Martínez[1], Julianna Juárez-Moya[1], Rogelio Sánchez Vega[1], Juan Manuel Tirado[1] and Raúl Alberto Reyes-Villagrana[1,2*]

1 Faculty of Animal Science and Ecology, Autonomous University of Chihuahua, Chihuahua, Chihuahua, Mexico

2 National Council of Science and Technology, Mexico City, Mexico

*Address all correspondence to: rareyesvi@conacyt.mx

References

[1] Darwin CR. On the Origin of Species by Means of Natural Selection, or Preservation of Favored Races in the Struggle for Life. 1st ed. London: John Murray; 1859

[2] Newton I. Philosophia Naturalis Principia Mathematica. 1st ed. London: Royal Society; 1686

[3] Maxwell JC. A Treatise on Electricity and Magnetism. 1st ed. Oxford: Clarendon Press; 1873

[4] Reyes-Villagrana RA, La Nanotecnología en nuestras manos: MEMs, 2do Congreso Nacional de Mecatrónica, Universidad Tecnológica del Estado de Zacatecas, Guadalupe, Zacatecas, México; 2007

[5] Reyes-Villagrana RA. Ondas acústicas, ultrasonido de alta intensidad y el efecto de la cavitación acústica, Seminario del C. A. Tecnología de Productos de Origen Animal, FZyE, UACh, Chihuahua, Chihuahua, México; octubre 2018

[6] Rossing TD, editor. Springer Handbook of Acoustics. 1st ed. New York: Springer; 2007

[7] Randall RH. An Introduction to Acoustics. USA: Dover; 2005

[8] Carrillo-López LM, Alarcón-Rojo AD, Luna-Rodríguez L, Reyes-Villagrana R. Mint: Modification of food systems by ultrasound. Journal of Food Quality. 2017;**2017**:5794931

[9] Born M, Wolf E. Principles of Optics: Electromagnetic Theory of Propagation, Interference and Diffraction of Light. USA: Cambridge University Press; 1999

[10] Mason TJ, Chemat F, Ashokkumar M. Mint: Power ultrasonics for food processing. In: Gallegos-Juárez JA, Graff KF, editors. Power Ultrasonics, Applications of High-Intensity Ultrasound. 2015. pp. 815-843

[11] Arnau E, editor. Piezoelectric Transducers and Applications. 2nd ed. Berlin: Springer-Verlag; 2008

[12] Szabo TL, Lewin PA. Mint: Piezoelectric materials for imaging. Journal of Ultrasound in Medicine. 2007;**26**:283-288

[13] Nye JF. Physical Properties of Crystals: Their Representation by Tensors and Matrices. UK: Oxford University Press; 1985

[14] Giraud F, Giraud-Audine C. Piezoelectric Actuators: Vector Control Method, Basic, Modelling and Mechatronic Design of Ultrasonic Devices. 1st ed. Oxford: Butterworth-Heinemann Elsevier; 2019

[15] Feng H, Yang W, Hielscher T. Mint: Power ultrasound. Food Science and Technology International. 2008;**14**:433-436

[16] Sistema de ultrasonido de alta intensidad, tipo sonotródo, Branson® Ultrasonic, Sonifier® SFX250/SFX550 Cell Distruptors. enero/2022

[17] Baño ultrasónico, Branson® Ultraosnics Brasonic® serie CPX. Enero/2022

[18] Franc J, Michel JM. Fundamentals of Cavitation. USA: Kluver Publishers; 2005

[19] Weissler A. Mint: Sonochemistry: The production of chemical changes with sound waves. The Journal of the Acoustical Society of America. 1953;**25**:651-657

[20] Crum LA. Mint: Sonoluminescene, sonochemistry, and sonophysics. Journal of Acoustic. 1994;**95**:559-562

[21] Crum LA. Mint: Comments on the evolving field of sonochemistry by a cavitation physicist. Ultrasonics Sonochemistry. 1995;**2**:S147-S152

[22] Suslick KS. Mint: Mechanochemistry and sonochemistry: Concluding remarks. 2014;**170**:411-422

[23] Rayleigh L. Mint: On the pressure developed in a liquid during the collapse of a spherical cavity. Philosophical Magazine Series. 1917;**6**:94-102

[24] Leighton TG. The Acoustic Bubbles. 1st ed. Cambridge: Elsevier Academic Press; 2012

[25] Gilmore, F. R., Mint: The Growth of Collapse of a Spherical Bubble in a Viscous Compressible Liquid. Report No 26-30, Hydrodynamics Laboratory, California Institute of Technology, Pasadena, California, USA, 1952, 1-41.

[26] Keller JB, Miksis M. Mint: Bubble oscillations of large amplitude. Journal of the Acoustic Society of America. 1980; **68**:628-633

[27] Suslick KS, Flannigan DJ. Mint: Inside a collapsing bubble: Sonoluminescence and the conditions during cavitation. Annual Review of Physical Chemistry. 2008;**59**: 659-683

[28] Suslick KS, Flint EB. Mint: The temperature of cavitation. Science. 1991; **253**:1397-1399

[29] Lauterborn W, Kurz T. Mint: Physics of bubble oscillations. Reports on Progress in Physics. 2010;**73**: 106501

[30] An Y. Mint: Formulation of multibubble cavitation. Physical Review E. 2011;**83**:066313

[31] Hebra AJ. The Physics of Metrology. New York: Springer; 2010

[32] Borisenko VE, Ossicini S. What is What in the Nanoworld. Germany: Wiley-VCH; 2008

[33] Feynmann RP. There's Plenty of Room at the Bottom. Pasadena, CA.: American Physical Society; 1959

[34] Chaudhry Q, Castle L, Watkins R, editors. Nanotechnologies in Foods. UK: RSC Publishing; 2010

[35] Boye JI, Arcand Y. Green Technologies in Food Production and Processing. New York: Springer; 2012

[36] Amalraj A, Gopi S, Thomas S, Haponiuk JT. Mint: Cellulose nanomaterials in biomedical, food, and nutraceutical applications: A review. Macromolecular Symposia. 2018;**380**: 1800115

[37] Gallocchio F, Belluco S, Ricci A. Mint: Nanotechnology and food: Brief over of the current scenario. Food Science. 2015;**5**:85-88

[38] Rahidi L, Khosravi-Darani K. Mint: The applications of nanotechnology in food industry. Reviews in Food Science and Nutritions. 2011;**51**:723-730

[39] Chau CF, Wu SH, Yen GC. Mint: The development of regulation for food nanotechnologies. Trends in Foods Science & Technologies. 2007;**18**:269-280

[40] Hamad AF, Han JH, Kim BC, Rather IA. Mint: The interwine of nanotechnology with the food industry. Saudi Journal of Biological Sciences. 2018;**25**:27-30

[41] Sattler KD, editor. Handbook of Nanophysics. USA: CRC Press; 2011

[42] Vasilescu A, Hayat A, Gáspar S, Marty JL. Mint: Advantages of carbon nanomaterials electrochemical apta sensors for food analysis. Electroanalysis. 2018;**30**:2-19

[43] Tripathi RM, Puddake RN, Shrivastav BR, Shrivastav A. Mint: Antibacterial activity of poly (vinyl alcohol) biogenetic silver nanocomposite film for food packaging material. Advanced Natural. 2018;**9**:025020

[44] Ayhan Z, Cimmino S, Esturk O, Duraccio D, Pezzuto M, Silvestre C. Mint: Development of films of novel polypropylene based nanomaterials for food packaging applications. Packaging Technology and Science. 2018;**28**: 289-602

[45] Pavlov YS, Revina AA, Souvorova OV, Voropaeva NL, Chekmar DV, Abkhalimov EV, et al. Mint: Electron beam agrobionanotecnologies for agriculture and food industry enable by electron accelerators. IOP Conference Series: Journals of Physics. 2017;**941**:012098

[46] Martirosyan A, Schneider YJ. Mint: Engineerd nanomaterials in food: Implications for food safety and consumer health. International Journal of Environmental Research and Public Health. 2014;**11**:5720-5750

[47] Balestra GM, Fortunati E, Giovanale G, Mazzaglia A. Mint: Innovation nanotechnological tools in plant and food protection. AIP Conference Proceedings. 1990;**2018**: 020017

[48] Gnoshinski IV, Khotimchenko SA. Mint: Nanomaterials in consumer's goods: The problems of risk assessment. IOP Conf. Series: Materials Science and Engineering. 2015;**98**:012009

[49] Aschberger K, Gottardo S, Amenta V, Arena M, Botelho Moniz F, Bauwneester P, et al. Petters, Mint: Nanomateriasl in food, current and future applications and regulatory aspects. Journal of Physics Conference Series. 2015;**617**:012032

[50] Tsagkaris AS, Tzegkas SG, Danezis GP. Mint: Nanomaterials in food packaging: State of the art analysis. Journal of Food Science and Technology. 2018;**55**:2862-2870

[51] He X, Hwang HM. Mint: Nanotechnology in food science: Functionality, applicability, and safety assessment. Journal of Food and Drug Analysis. 2016;**24**:671-681

[52] Padua GW, Wang Q. Nanotechnology Research Methods for Foods and Bioproducts. India: Wiley-Blackwell; 2012

[53] Wyser Y, Adams M, Avella M, Carlander D, García L, Pieper G, et al. Mint: Outlook and challenges of nanotechnologies for food packaging. Packaging Technology and Science. 2016;**29**:615-648

[54] Bajpai VK, Kamle M, Shukla S, Mahato DK, Chandra P, Hwang SK. Mint: Prospect of using nanotechnology for food preservation, safety, and security. Journal of Food and Drug Analysis. 2018;**26**:201-214

[55] Dimitrijevic M, Karabasil N, Boskovic M, Teodorovic V, Vasilev D, Djordjevic V, et al. Mint: Safety aspects of nanotechnology applications in food packaging. Procedia Food Science. 2015; **5**:57-60

[56] Rossi M, Cubaddad F, Dini L, Terranova ML, Aureli F, Sorbo A, et al.

Mint: Scientific basis of nanotechnology, implications for the food sector and future trends. Trends in Food Science and Technology. 2014;**40**:127-148

[57] Chellaram C, Murugaboopathim G, John AA, Sivakumar R, Ganesan S, Krithika S, et al. Mint: Significance of nanotechnology in food industry. APCBEE Procedia. 2014;**8**:109-113

[58] Carbone M, Tommasa Donia D, Sabbatella G, Antiochia R. Mint: Silver nanoparticles in polymeric matrices for fresh food packging. Journal of King South University Science. 2016;**28**:273-279

[59] Bhat S, Tuinier R, Schurtenberger P. Mint: Sponodal decomposition in a food colloid biopolymer mixture: Evidence for a linear regime. Journal of Physics. 2006; **18**:339-346

[60] Lizhao H. Mint: The analysis about the application of nanomaterial in food packaging. IOP Conf. Series: Materials Science and Engineering. 2018;**439**: 042021

[61] Awad TS, Moharram HA, Shaltout OE, Asker D, Youssef MM. Mint: Applications of ultrasound in analysis, processing and quality control of food: A review. Food Research International. 2012;**48**:410-427

[62] Abbas S, Hayat K, Karangwa E, Basshari M, Zhang X. Mint: An overview of ultrasound-assisted food-grade nanoemulsions. Foods Engineering Review. 2013;**5**:139-157

[63] Souza VGL, Fernando AL. Mint: Nanoparticles in food packaging: Biodergradability and potential migration to food a review. Food Packaging and Shelf Life. 2016;**8**:63-70

[64] Parakoti K, Manibolu M, Hwang HM. Mint: Nanostructures: Current uses future applications in food science. Journal of Food and Drug Analysis. 2017;**25**:245-253

[65] Bouwmeester H, Van der Zande M, Jepson A. Mint: Effects of food-borne nanomaterials on gastrointenstinal tissue and microbiota. WIRES Nanomd Nanobiotechnology. 2018;**10**:1481

[66] Bumbudsanpharoke N, Ko S. Mint: Nano-food packaging: And overview of market, migration research, and safety regulations. Journal of Food Science. 2015;**80**:910-923

[67] Taurozzi JS, Hackley VA, Wiesner MR. Mint: Ultrasonic dispersion of nanoparticles for environmental, health and safety assessment—issues and recommendations. Nanotoxicology. 2011;**5**:711-729

Chapter 3

Nutrigenomics: Challenges and Opportunities

Aliza Batool, Umar Farooq, Afshan Shafi and Zulqurnain Khan

Abstract

Nutritional genomics is one of the emerging fields of food sciences for innovative trends in food sciences. Understanding of the genetics of the human health and diseases is very important to set the diet and nutrition plans. Functional genomics studies have paved the path to the cure of the disease with diet. With the advancement in the field of genetics and genomics especially next generation sequencing and molecular markers, nutrigenomics has been gaining much attention in the field of food sciences. The chapter will elaborate challenges and opportunities associated with the field of nutrigenomics and will propose strategies to address the issues.

Keywords: nutrition, genetics, genomics, diseases, heath

1. Introduction

Nutrigenomics is a discipline of nutrition that uses molecular approaches to discover, access, and understand the varied reactions produced by a certain diet in individuals or communities. Its goal is to figure out how the components of a certain diet (bioactive chemical) affect gene expression, which might have increased or decreased its potential [1].

Nutritional genomics (or nutrigenomics) as the field of study that studies all forms of interactions between food and the genome and is characterized by the use of high-throughput genomic (or genome related) methods. Nutritional genetics (also known as nutrigenetics) is a branch of nutrigenomics that studies how genomic variants interact with dietary variables and the consequences of these interactions [2].

Nutrigenomics is used to integrate genetic methods into nutrition studies. Nutrigenomics (also known as nutritional genomics) is the study of the connections between meals or dietary supplements and an individual's genome, as well as the phenotypic impacts of these interactions [3].

2. Basics of nutrigenomics

Diet play an important role on the health of human. However nutrigenomics is the interaction of food-gene interactions, sometimes known as 'inborn errors of metabolism,' which have long been corrected by dietary changes [4] (**Figure 1**). Nutrigenomics is a relatively recent science that explains how some foods affect your

Figure 1.
Interaction of nutrient and gene.

DNA. What you eat has a direct impact on the genetic messages your body receives. It is significantly change how food interacts with your body, reduce weight, and enhance your health if you can learn the language of your genes and regulate the messages and instructions they provide your body and metabolism." The study of the impact of nutrition on the genome, proteome, and metabolome is known as nutrigenomics [5].

Nutrigenomics tries to determine the impacts of various nutrients on the genome, including macronutrients and micronutrients, and investigates the interactions between genes and nutrients or food bio-actives, as well as their consequences on human health. Nutrigenomics is the study of the effects of nutrition on transcription varied responses of genomic variations and gene expression. Nutrigenomics deals with study of a biological system using functional genomic methods in order to learn how dietary components alter metabolic pathways as well as homeostatic control [6]. Nutrigenetics will provide crucial information to help clinicians decide the ideal food for a specific person, i.e. individualized diet. Nutrigenomics research employs tools such as transcriptomic, proteomics, and metabolomics. Nutrition has been shown to alter expression of genes on numerous levels, chromatin structure, including gene function, signal transduction and protein function [5]. Food intake has been linked to the prevalence and severity of chronic diseases. Type 2 diabetes, obesity, metabolic syndromes, cardiovascular disease, and some kind of cancer are among the many nutrition-related illnesses that are polygenic and multifactorial, with onset and the progression linked to several s genes variations, as well as a variety of environmental factors, including diet [7].

Nutrigenomics is a new and emerging discipline of genomics that studies the impact of dietary consumption on the whole genome (full genetic make-up, including epigenetic modifications), proteome is the total number of proteins, and metabolome is the whole number of molecules in the body and the sum of all metabolites. Galactosemia, for example, is a disease caused by an inherited genetic deficit in one of the three enzymes involved in galactose processing [8].

3. Genomics: a molecular approach

Functional genomics technologies like metabolomics, transcriptomics and proteomics have a few things in common. For starters, they are non-targeted, all-encompassing technology. Second, they rely on determining the actual concentrations of a single type of biomolecule's entire set for example protein, mRNA or metabolite. Finally, they associate the entire collection of particular type of biomolecule (transcriptomes, proteomes, or metabolomes) under various environmental situations [9].

Above discuss technologies has different purpose. When investigating the biological effects of functional food bio actives, for example, a transcriptomics technique appears the most suited since it provides a comprehensive overview of a cell's reaction to a bioactive that can be compared to known "healthy" or "unhealthy" responses. A metabolomics method is very useful in metabolic engineering because the metabolome is the closest to the product/phenotype under examination [9]. The fundamental benefit of an applied genomics method is that it allows scientists for the first time to get a comprehensive understanding of the biological activities that happen in cells in response to specific therapies. We will obtain a greater knowledge of these processes, allowing us to direct them in the right direction for example, by slowing down or even inhibiting disease development. Obtaining massive amounts of data is no longer a problem, but interpreting that data (converting data into knowledge) remains a substantial bottleneck. This is due in part to the fact that a large number of genes have yet to be assigned a function, making biological data interpretation on these genes challenging. Additional, and perhaps the most pressing issue at the moment, is that researchers in the life sciences have never been taught how to deal with vast volumes of data and interpret them holistically, but have instead been taught to employ a reductionist approach. Recognizing this constraint, on the other hand, is the first step toward finding a solution [10].

3.1 Transcriptomics

Transcriptomics is the study of the entire set of RNA transcripts that have been activated. Because mRNAs are synthesized at a specific time, in specific tissue of a specific organism, expression of gene changes depending on the conditions and time periods [11]. When activated, transcription factors travel toward nucleus and attach to a particular DNA sequence found in the promoter region of genes, where they block or facilitate transcription [12]. Transcriptomics can help to provide data of mechanisms or underlying impacts of a certain nutrient or diet in nutrition studies. It help in recognizing and characterizing the pathways controlled through bioactive compounds and nutrients in foods, as well as it identify the metabolites, genes and proteins that modify as predeceases progress [1].

3.2 Proteomics

Proteomics is the molecular approach is used to investigate the entire form of proteins complex in a species biological activities. These proteins function normally in the tissue, organ, and cell but in particular biological conditions, such as transcriptome, and they can change their activity and the level of gene expression. Proteins are a type of molecule that can be found in every living cell [13]. They have structural, transport, biochemical, cell signaling, mechanical and storage functions in the cell. They're also a necessary component of the human diet. The amount of proteins made by an organism is higher as compared to the number of genes it has. This occurs as a result of many posttranslational and translational changes [14]. Proteomics is the study of protein expression using a variety of technologies. It does so by employing techniques like chromatographic electrophoresis, prefractionation of samples using extraction sequences, and organellar proteome analysis, among others. As a result, Proteomics is a critical resource for Nutrigenomics, as it establishes a connection between genome sequences and cell behavior, and acts as a biological tool for understanding the process of genetic function assessment, as well as how the genome is activated in response to a certain diet. The butyrate activities change the expression of many proteins in the ubiquitin-proteasome system. Butyrate modulates essential proteins involved in the process of apoptosis, cell differentiation and cell cycle [1].

3.3 Metabolomics

The metabolome of an organism or species is made up of a collection of tiny secondary and primary metabolites as well as bodily fluids. Metabolomics is a branch of functional genomics that analyses metabolite alterations with the goal of isolating and characterizing them. Advances in this field may make it easier to comprehend how a person's genotype influences his or her phenotype [15]. Metabolomics has various nutritional uses, one of which is to determine the patterns and metabolic problems induced by a human food, as well as how these changes may affect the health or sickness of human As a result, metabolomics analyses metabolism in the context of genetic and environmental disturbances, which may be examined and interpreted using bioinformatics and statistical methods [16]. Small chemical substances that directly interact with proteins and other macromolecules and are dissolved in the cell cytoplasm are referred to as metabolites. There are two types of metabolites: main and secondary. Primary metabolites are directly engaged in macromolecule synthesis and breakdown, whereas secondary metabolites are more abundant in plants and serve as an important compound and defense mechanisms [17]. In the field of nutrition, metabolomics enables researchers to better understand metabolic instabilities and arrangements that are caused or exacerbated by diet. This advances our understanding of how an individual's health/illness is affected by an extra or absence of certain nutrient or molecules (secondary metabolites) contained in food. These chemicals (nutrients or not) interact with the body in a variety of ways, altering metabolome pathways. Perilla alcohol (a monoterpene derived from strawberries) can, for example, act as an anticancer molecule when exposed to particular chemical stimuli [1].

4. Diet-gene-disease triangle

Dietary substances can have a direct or indirect effect on gene expression. Nutrients could act as molecules for transcription factor sites or to be digested via primary or secondary biosynthetic processes at the cellular level, changing substrate or intermediate quantities and so favorably or unfavorably modulating signal pathways. Among the most essential molecules involved in nutrient-induced gene activity is transcription factors (TFs). The PPARs TFs, which include 48 individuals in the genetic material, are among the most important groups of nutrition sensors [18]. The preponderance of sensors in this family connect nutrients and their metabolites, as well as regulate gene expression in fatty acid gluconeogenesis, amino acid metabolism, oxidation, ketogenesis, cellular proliferation and the acute-phase response of the liver. PPAR-ligands include the fatty acids arachidonic acid (20:4n6), palmitic (16:0), linoleic (18:2n6) and oleic (18,1n9), as well as the 8- (S) hydroxyeicosatraenoic acid, eicosanoids 15deoxy-12 and 14prostaglandinJ2. Fatty acid sensors are provided by these nuclear receptors. Hyperforin binds to nuclear receptors directly and regulates genes. Lipid sensors typically form heterodimers with retinoid receptors, which have a ligand derived from vitamin A, another dietary nutrient [10].

The liver X receptor binds to certain nucleotide sequences (response elements) in the promoter regions of numerous genes as heteromers. In response to ligand engagement, nuclear receptors change their shape, allowing corepressors to dissociate and coactivator proteins to connect in preparation for transcriptional activation. The key metabolic condition activates a variety of genes, such as those implicated in fatty acid oxidation & storage. These TFs operate as nutrition sensors in metabolically active organs like the liver, stomach, as well as adipose tissue via changing the degree of DNA transcription of particular genes in response to food changes [3]. Some TFs are regulated indirectly by dietary substances. Protease cleavage, for example, activates sterol regulatory element binding proteins (SREBPs), minimal quantities of oxy sterols, insulin/glucose, and polyunsaturated fatty acids govern this process. High glucose levels activate the carbohydrate-responsive element-binding protein (chREBP), which is controlled via reversible phosphorylation mechanisms. The DNA-binding protein regulates the expression of lipogenic genes. Furthermore, dietary substances have the ability to influence signal transduction pathways directly. Green tea, for example, includes the polyphenol 11-epigallocatechin-3-gallate (EGCG), which inhibits tyrosine phosphorylation of the Her-2/neu receptor and the epidermal growth factor receptor, reducing signaling via the PI-3-AKt kinase-NF-kB pathway. The stimulation of the NF-kB cascade has indeed been associated toward certain types of breast cancer [4].

Other nutrients, including such polyunsaturated fatty acids (PUFAs) like n-3 and n-6, often known as omega-3 and omega-6 fatty acids, were shown to influence gene expression. Several proteins gene expression are involved in lipid and glucose metabolism has been found to be affected by PUFA ingestion in animal studies. There has also been evidence of a link between the PPARA Lue162Val polymorphism and n-6 PUFA intake. In those who hold the less common V162 allele, increased n-6 PUFA intake is connected to a considerable decline in triacylglycerol concentration, though not in people who carry the L162 variety. Conversely, n-3 PUFA consumption reduces triacylglycerol concentrations in L162 and V162 carriers. In the human diet, approximately 40 micronutrients are required. CVD has been linked to inadequate intakes of key micronutrients. Nutrient deficiencies are much more important than radiation due to the obvious regularity of exposure to an environment that causes DNA damage. The large presence of uracil in human DNA (4 million uracil/cell) causes chromosomal disruption in folate deprivation. Nutrients like folate, carotenoids, Vitamin D, B6, B12 helpful for the management of cancer, neural tube abnormalities. These nutrients help in Hyperhomocysteinemia that is major risk for cardiac disease. Vitamin B12, niacin, folic acid, vitamin C, vitamin E, zinc, iron and B6 deficiency appears to damage DNA in the same way that radiation does, causing single and double-strand breaks, oxidative lesions, or both [10].

Amino acids can act as dietary signals by modulating the expression of specific genes. Cells have been shown to detect variations in amino acid levels and respond through mechanisms such as transcription control, mRNA stability, and up or down regulation of translation start. Blood glucose levels are influenced differently by simple and complex carbohydrates. High glycemic index (GI) food produce more insulin while producing fewer insulin receptors. Several genes involved in the glycolytic and lipogenic pathways are transcribed when blood glucose levels are high. Food-regulated genes must be implicated in the beginning, progression, and severity of disease since dietary components are regularly eaten and are involved in the control of gene expression both indirectly and directly [7].

5. Life style associated metabolic diseases

The main factor related with today's lifestyle is unhealthy living habits. It is culminating in diseases with high mortality rates, particularly chronic diseases, which are responsible for the majority of fatalities over the last decade. Lifestyle-related diseases are a set of diseases caused by humankind's long-term exposure to bad diet, lifestyle, and living conditions. Cardiac, renal failure, nutrition-induced malignancies, hypertension, diabetes, chronic bronchitis and other diseases have nearly identical risk factors because they are slow-progressing, non-infectious, and non-transmissible. The primary motivation for choosing lifestyle-related disorders for this assessment is their impact on human health. According to the WHO research, illness profiles have shifted rapidly from communicable to non-communicable diseases during the last few decades, regardless of area, race, or economic status. Around 60% of deaths globally were due to lifestyle-related chronic diseases, which were twice as common as infectious diseases. Non communicable diseases were also responsible for 53 percent of deaths in India, with cardiovascular disease accounting for 24 percent of all deaths (CVDs). These disorders, like epidemiological characteristics, have particular metabolic risk factors related with physiological mechanisms that result in mitochondrial changes, oxidative stress, and inflammation. These reactions to changes in the environment play an important role in the onset and progression of lifestyle disorders. The increase in metabolic risk factors of blood such as blood pressure, glucose, lipids, and other variables is primarily due to a poor diet. Inflammation in the human body is also affected by diseases related with modern lifestyles. Eicosanoids (arachidonic, eicosapentaenoic and docosahexaenoic acid), a fatty acid metabolite, regulate the inflammation process. Inflammation has been linked to a poor diet and the current social and environmental stressors that humans encounter. In an animal model, scientists discovered the genes (plasminogen activator inhibitor-1 linked with fat) that were responsible for the change. Changes in lifestyle and environment have an impact on not just human metabolic and physiology processes, but also the intestinal microbiome, which can lead to health issues. The idea that these diseases are only found in rich countries has been debunked, as low- and middle-income countries have more favorable social, economic, and environmental conditions for their development. If this change continues, illness profiles will worsen, especially in developing and underdeveloped countries [19].

6. Nutrigenomics and chronic diseases

Nutrition is the act of providing various constituents to the organism. Nutrients play important functions carbohydrates and fats are the source of energy, for the structure of cell protein is the best sources, vitamins and minerals is good for the control of metabolism, allowing the organism to maintain its homeostasis. The combination of different factors, like social state, emotional, physical activity and genetic background, determines an individual's nutritional status [20]. Diet plays an important role because the minerals and other bioactive chemicals found in food can either be healthy or cause a variety of disorders. Many chronic disease like phenylketonuria, cancer, diabetes, and dyslipidemias, are among the disorders linked to food consumption. In this approach, a person's health will be determined by the interaction of their genes and their dietary habits. As a result, nutrigenomics, like other omic sciences, aims to better understand the relationship among genes and diet (nutrients) [21].

6.1 Obesity

Obesity is the common problem now days different factors are responsible like environmental, genetic factors, lifestyle and metabolic syndromes. Genetic variables account for 80 percent of the variance in body mass index (BMI) between twins. Because obesity generates a persistent inflammatory response, using Nutrigenomics to control it is extremely promising. In food bioactive compounds are present like in olive oils tyrosol present, in fruits and greeneries quercetin present and in tomatoes lycopene is present etc. [22].

Food-derived bioactive chemicals can also interfere with genes in different ways. One of the principal pathways for the expression of gene modification is during transcription, when inflammatory mediators are synthesized, which plays a key role in a variety of chronic diseases, including obesity. Interleukin-1 is one of these mediators, and it induces the creation of numerous other molecules during the inflammatory cascade after it is activated. Green tea bioactive ingredient tocopherol works as an anti-inflammatory [12].

6.2 Cancer

Different factors are responsible for cancer like age, genetic, lifestyle, physical activity, and diet. Due to deficiencies of nutrients like vitamins E, C, B12, as folic acid, vitamin B6, C, and as well as zinc, selenium and niacin, have previously been found to cause change in DNA similar to those seen after contact to radiation. These changes can result in double-strand DNA breakage and oxidative damage. They were also found to be closely linked to the growth of cancer. Contaminated food produce harmful metabolites that regulate the gene expression [13].

Mutation has the potential to cause serious liver problems such cancer, necrosis and cirrhosis. During metabolism of folate, folic acid in food is absorbed by the intestine and converted into 5-methyltetrahydrofolate by a series of chemical catabolism and synthesis processes. 5-methyltetrahydrofolate component is required for the production of methionine, which is employed in the DNA mutilation process. Therefore, a low-folic acid diet can disrupt this process and also interfere with the replication of DNA, increasing the risk of cancer formation. Nonetheless, many minerals defend against the development of cancer [23]. Selenium stimulates the production of glutathione peroxidase, an enzyme that reduces hydrogen peroxide and maintains the integrity of cell membranes and zinc which also influences processes such as apoptosis genomic stability modulation and genetic expression [24].

6.3 Diabetes

Around 90% of people in world suffer from diabetes. Type 2 diabetes is a complex disease that is caused by a combination of hereditary and environmental factors. There are Single-nucleotide polymorphism SNPs linked to the likelihood of acquiring type 2 diabetes. Tests for the detection of SNPs linked to Type II Diabetes became available to the general public as genome sequencing and decoding progressed. During these tests, the patient might find out if he or she has a genetic predisposition to developing the condition. Due to low food income and insulin tolerance, some people got diabetes later in life. Patients who had a positive result for the existence of type II diabetes changed their lifestyles, particularly their eating habits, which later reduced the disease's progression in this group [1].

6.4 Cardiovascular

CVD is defined by the creation of intimal lesions in the blood arteries as a result of fibrosis, lipid buildup, inflammatory response, and cell death. According to a

WHO report, cardiovascular disease is the leading cause of death. Global estimates put the death toll at 17.5 million in 2012, representing for 31% among all deaths. In the treatment and prevention of cardiovascular disease, diet is essential. Changes in diet can protect genes involved in lipid metabolism and synthesis, such as arachidonate 5-lipoxygenase (ALOX5), apolipoprotein E (APOE), fatty acid synthase (FASN), and peroxisome proliferator activated receptors (PPARs), lipoprotein lipase (LPL), and others. The 28.1-kDa protein apolipoprotein (APOA1) is important for bodily disposal and is a form of high lipoprotein (HDL). This gene is used as a biomarker to diagnose cardiovascular illnesses like myocardial infarction in the early stages. As a result, HDL clearance from plasma is necessary while eating a low-fat diet. PUFAs play a key function in influencing the expression of factors involved in glucose and lipid metabolism [25]. Fatty acids helps individuals lower their LDL cholesterol levels. Lipoxygenase, also known as arachidonate 5-lipoxigenase (5-LOX), is an important enzyme that controls the production of leukotrienes, inflammatory cytokines, and chemokines. Its expression is increased in atherosclerotic lesions, causing more inflammatory cells to be mobilized. Endothelial nitric oxide synthase in blood arteries converts L-arginine to nitric oxide (NO). By activating the guanylyl cyclase receptor and boosting cGMP levels, NO migration into vascular smooth cells devalues blood cells. It also works as a leukotriene and platelet aggregation inhibitor, lowering the risk of white blood cell adhesion and atherosclerosis. Increased eNOS expression has been linked to the creation of H2O2 and has been linked to endothelial dysfunction, which leads to CVD. In both animals and humans, L-arginine has been demonstrated to be useful in the treatment of hypercholesterolemia. Omega three polyunsaturated fatty acids increase eNOS expression, resulting in vasorelaxation, while lowering circulating indicators such as E-selectin, intercellular adhesion molecule-1, and vascular cell adhesion molecule-1 (VCAMs- 11) [26].

6.5 Cholesterol

The endothelium and leukocytes interact through the vascular cell adhesion molecule-1 (VCAM-1). During the start of atherosclerosis VCAM-1 binds monocytes and T lymphocytes, resulting in plaque development. However, proinflammatory cytokines such as tumor necrosis factor- (TNF-) and interleukin-1 generate oxidized lipids necessary for VCAM-1 expression (IL-1) [19].

6.6 Inflammation

The inflammatory process is known to be influenced by the nutritional composition of diet systems resulting in immune response regulation, via altering gene expression as well as interfering with signaling cascade. Inflammation is the body's biological response to any invasion or injury, and it can be acute (short-term self-healing) or chronic (long-term). Inflammation later in life might linger for many years, leading to a variety of disorders such as cancer, intestinal allergies, atherosclerosis, and so on [23]. The immune system regulates the inflammatory process, and numerous immunological mediators play a key role. The famous spice curcuma longa (turmeric) has long been used in Ayurvedic medicine to treat inflammation. It includes a variety of bioactive chemicals, but 'curcumin' is one of the most important metabolites, accounting for a wide range of pharmacological actions including antioxidant, antibacterial, and anti-inflammatory effects. There is evidence that curcumin is a greatly pleiotropic chemical having potential to interact with inflammatory molecular targets. This was discovered as promising

Nutrients	Gene	Diseases	Reference
Curcumin	H+, K + - ATPase	Help in the treatment of gastric ulcers	[27]
	IL-6,TNF-_, COX2	Reduces inflammation	[28]
	SOCS1–3,p38MAPK	Trans locates nuclear material in a different way.	
	TLRs & MyD88	Infection with Helicobacter pylori has an anti-inflammatory impact.	[29]
Retinoic Acid	TLR4/NF-kappa B	It helps to minimize mastitis irritation.	[30]
Vitamin E	CRP, IL-8, PAI_1	Arthrosclerosis	[31]
Sesamol	MCP-1, RANTES, IL-1_, IL-6, and CXCL-16	Reduce the number of lesions caused by arthrosclerosis.	[32]
	ABCA1, ABCA2, APO2, LCAY, CYP7AI, APOE,	Help in cholesterol metabolism and transportation	[32]
Vitamin B6	TNF_,, COX-2, iNOS and IL-6, IFN-_	Help in the inflammatory bowel diseases	[33]
Grains, legumes and refined rice	APOA5 c.-1131CNT and triglyceride and APOA-V	Diabetes type 2	[34]
Polymorphisms	APOA5	Uyghur	[35]
polymorphism	Triglycerides and the APOA5 c.-1131TNC,	Age-related association in mice	[35]
	APOC3 (rs5128)	Metabolic syndrome	[36]
	1β (PGC-1β), a co-activator of PPARγ	It reacts to changes in the macronutrient content of the food.	[37]
Selenoproteins	DIO1, SEPHS1, DIO2, GPX1, TXNRD2, GPX3 and SEPSECS	Crohn's disease	[38]
Zinc	ZNF365	Low zinc levels	[39]
Eicosapentaenoic acid	omega-3 (n − 3) fatty acids are docosahexaenoic acid (DHA) the long chain polyunsaturated	Inflammation	[40]
Vitamin D	NOD2	Crohn's disease	[41]
Flavonoids and carotenoids	ω − 3 fatty acids	Gene expression and oxidative stress production of inflammatory mediators	[42]
Gluten	HLA-DQ (DQ2 and/or DQ8)	Coeliac disease	[43]
SNPs.	SNPs	Obesity, diabetes, cancers, cardiovascular disease and brain disorders	[44]
Avian responses to nutrition	IGF1 and lysine (K)-specific histone demethylase 5A (KDM5A)	Growth and epigenetic regulation of genes	[45]
Vitamin D	Retinoic acid receptor (RAR) binds	Nutritional imbalances have been linked to aging, behavioral problems, chronic fatigue, cardiovascular disease (CVD), diabetes, neurological disorders, cancer, immunological diseases, stroke, multiple sclerosis, Parkinson's disease.	[46]

Nutrients	Gene	Diseases	Reference
Folate-related enzyme	677C → T	Hyperhomocysteinemia	[47]
SNPs	Glucose-6-phosphate dehydrogenize (G6PD) gene, Phenylalanine hydroxylase gene and galactose-1-phosphate uridyltransferase (GALT) gene	Prevent from disease Phenylketonuria (PKU), Favism and Galactosemia,	[48, 49]
Selenium	Selenium-dependent Enzyme	Lung cancer	[50]

Table 1.
Nutrients gene interaction for the management of disease.

helpful not just for inflammatory diseases, but also for some tumors. Curcumin play an important role in the activity of H+, K + -ATPase and it also decreased the histone H3 acetylation, which hindered the expression of gene (H+, K + -ATPase). It regulates the production of cytokine signaling-1 (SOCS-1 and 3) and p38 MAPK suppressors, which suppress the expression of LPS-induced inflammatory mediators such as IL-6, TNF-_, and COX-2 mRNAs in macrophages. LPS-induced SOCS-1 and -3 expression as well as p38 MAPK kinase activation are reduced when nuclear translocation is altered further. TNF-_ activity and production are reduced in vitro, in vivo, and in people when curcumin (diferyloylmethane) is taken orally, according to bioavailability and safety studies. Vitamin A supplementation has been shown to help with a variety of inflammatory illnesses, including skin disorders and precancerous and cancerous situations. Retinoid, a vitamin A derivative, has been shown to prevent tumors growth and boost the immune system (**Table 1**) [19].

7. Nutrigenomics approach to functional food

Because of their interaction with diseases, nutrition and health are at the forefront of the scientific community, paving the way for the development of health-promoting food prototypes (s). Functional food is any food or food component that is beneficial for health and also provide essential nutrients. Functional component derived from plant like omega 3 fatty acid, antioxidant, vitamin, protein, dietary fibers, amino acid, flavonoids and prebiotics these compound help in the management of lifestyle related disease. Curcumin, a dietary flavonoid, inhibited adipogenesis in pre-adipocytes from humans. Curcumin reduce hyperglycemia and increase insulin sensitivity while also lowering TNF- levels. Oral treatment of Trigonella foenum graecum seed extract reduce the triglycerides, blood glucose, total cholesterol and increase high density lipoprotein. Due to presence of bioactive compound like 4-hydroxyisoleucine (amino acid) and the steroid saponin trigonelline fenugreek seed powder has hypoglycemic properties [51]. Functional compound has best therapy for the lifestyle related diseases. Similarly, omega-3 fatty acids DHA and EPA generated from the oil of fish have an important role in a variety of diseases, including cancer and cardiovascular disease. Supplements containing conjugated linoleic acid (CLA) have powerful antiobesity and hypolipidemic properties. Linoleic acid therapy was observed to increase body protein levels while at the same time lowering body fat levels. CLA supplementation increases adipocyte lipolysis while decreasing fat accumulation. Probiotics are a type of fermented food that promotes good health. Lactobacillus acidophilus have a hypocholesterolemic impact by suppressing the activity of 3-hydroxy-3-methylglutaryl CoA reductase, it is a crucial enzyme in the production cholesterol (**Figures 2** and **3**) [52].

7.1 Nutraceuticals

The human lifestyle has improved dramatically over the last few decades, but it has made some new pals in the shape of "lifestyle linked metabolic disorders." The main suspect in the nutritional disaster that led to the current condition of health is erroneous eating habits. Stephen DeFelice invented the word nutraceutical in 1989, defining it as "a food and any component of food beneficial for the treatment and management of human health [53]. Many phytochemicals like antioxidant, glycosinolates, carotenoids, flavonoids, phytoestrogens and dietary fiber are the most common use nutraceuticals. Bioactive compounds are presents in plants naturally show good response to human health like flavanols are present in green tea, resveratrol in grape seeds, lycopene in tomatoes and anthocyanins in blueberries etc. Flavonoids have an inverse connection between dietary intake and the risk of diabetes and cardiovascular disease. In human umbilical vein endothelial cells (HUVEC) stimulated by angiotensin II, the epigallocatechin-3-O-gallate (EGCG) flavanol present in green tea inhibited the protein and mRNA expression

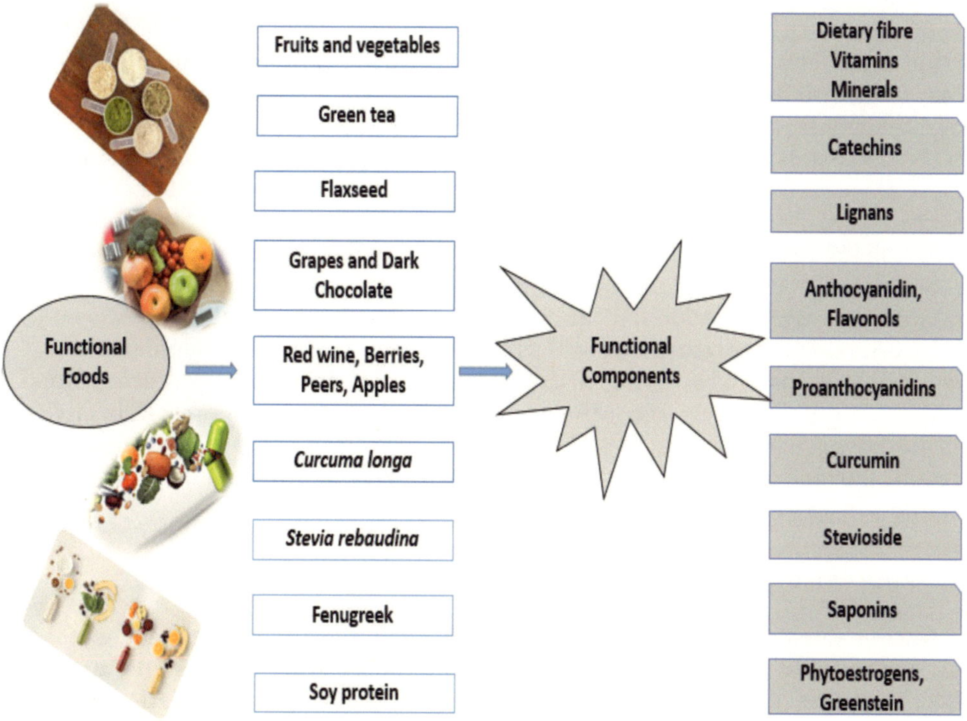

Figure 2.
Functional foods and its bioactive component.

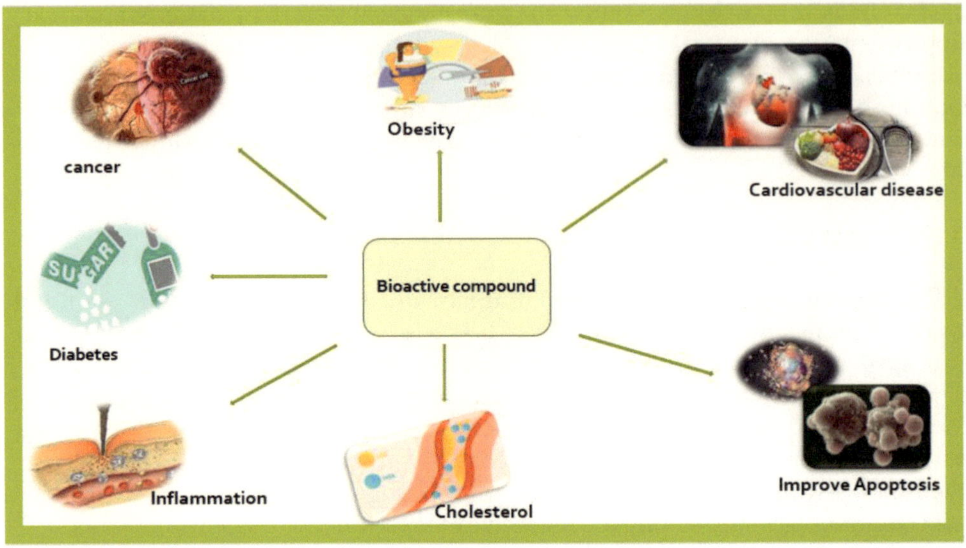

Figure 3.
Bio-active compound and their role in life style related disease.

of VCAM-1 and ICAM1 genes in a concentration-dependent manner (10 to 50 _M). Later research discovered a link between flavan-3-ol consumption and a lower risk of coronary heart disease mortality. Phloridzin, a dihydrochalcone, has also been shown to play a function in the treatment of postprandial hyperglycemia [54]. The use of dihydrochalcone and quercetin at various doses lower the level of blood sugar in type 2 diabetes mice and also quercetin supplementation lower the cholesterol level. In liver tissue dietary quercetin supplementation increase the activity of many antioxidant enzymes like glutathione peroxidase (GSH-Px), superoxide dismutase (SOD) and catalase (CAT). Quercetin supplementation lower the expression of PPAR it is used for hyperglycemia, hyperinsulinemia, and dyslipidemia. The use of naringenin-7-O-glucoside in a dose dependent manner improved the cardiac H9C2 cell line. Polyphenols found in lemon, such as flavanone, were found to protect against obesity caused by a high-fat diet. Lemon is the rich source of polyphenols like naringin, eriocitrin, and hesperidin, has higher levels of PPARs and acyl-CoA oxidase expression, as well as a lower body weight. The supplemented diet decreased insulin resistance in experimental animal models by altering blood insulin and glucose levels, according to the findings. Polyphenol present in citrus has vasculoprotective properties, which is very beneficial for cardiac patients. As a result, nutraceuticals have the ability to control and/or prevent nutrition-related disorders in addition to providing nutrients [55].

7.2 Future perspective

Nutrigenomics is the application of high-throughput genomics technology to nutrition studies. If it is used correctly, it will promote a better understanding the effect of diet on homeostatic control and metabolic pathway, the regulation is disrupted in initial stages of a diseases related to diet. In the future, nutrigenomics will enable efficient dietary intervention to restore normal homeostasis as well as prohibit diet-related diseases.

New commercial representations for individualized nutritional advising established on a person's DNA were also highlighted as a result of the knowledge integrating technology into health sciences. In this approach, organizations concerning the health-care system must act to control these commercial representations in order to protect patient's honesty while also improving the organization's nutrition focus performance.

8. Conclusion

Nutrigenomics and nutrigenetics are emerging disciplines that are expanding a variety of fields of study, with medicine, heredities, and diet. Population studies, large sample sizes, with proper experimental designs, scientific trials, and product-specific trials in people selected for specific genetic variations must now be empowered and examined using the massive amount of data obtained by GWAS. The ultimate purpose is to make more effective dietary intervention for individual techniques for preventative medicine and for the better quality of healthy life. Indeed, knowing very relevant to modulations in gene and diet could assist to profile the general dietary recommendations for each population from a Public Health perspective.

Author details

Aliza Batool[1], Umar Farooq[1], Afshan Shafi[1] and Zulqurnain Khan[2]*

1 Department of Food Sciences and Technology, MNS University of Agriculture, Multan, Pakistan

2 Institute of Plant Breeding and Biotechnology, MNS University of Agriculture, Multan, Pakistan

*Address all correspondence to: zulqurnain.khan@mnsuam.edu.pk

References

[1] Sales NMR, Pelegrini PB, Goersch M. Nutrigenomics: Definitions and advances of this new science. Journal of Nutrition and Metabolism. 2014;**2014**:1-6

[2] Mathers JC. Nutrigenomics in the modern era. Proceedings of the Nutrition Society. 2017;**76**(3):265-275

[3] Ferguson LR, Philpott M, Barnett MP. Nutrigenomics: Integrating genomic approaches into nutrition research. Molecular Diagnostics. 2010. pp. 347-363

[4] Bull C, Fenech M. Genome health nutrigenomics and nutrigenetics: Nutritional requirements for chromosomal stability and telomere maintenance at the individual level. The Proceedings of the Nutrition Society. 2008;**67**(2):146-156

[5] Neeha VS, Kinth P. Nutrigenomics research: A review. Journal of Food Science and Technology. 2013;**50**(3):415-428

[6] Fenech M. Genome health nutrigenomics and nutrigenetics-diagnosis and nutritional treatment of genome damage on an individual basis. Food and Chemical Toxicology. 2008;**46**(4):1365-1370

[7] Farhud DD, Yeganeh MZ, Yeganeh MZ. Nutrigenomics and nutrigenetics. Iranian Journal of Public Health. 2010;**39**(4):1

[8] Sharma P, Dwivedi S. Nutrigenomics and nutrigenetics: New insight in disease prevention and cure. Indian Journal of Clinical Biochemistry. 2017;**32**(4):371-373

[9] Trujillo E, Davis C, Milner J. Nutrigenomics, proteomics, metabolomics, and the practice of dietetics. Journal of the American Dietetic Association. 2006;**106**(3):403-413

[10] Bull Barnett MPG, Ferguson LR. Nutrigenomics: Integrating genomic approaches into nutrition research. In: *Molecular Diagnostics*. Cambridge, Massachusetts, USA: Academic Press; 2017. pp. 305-326

[11] Fialho E, Moreno FS, Ong TPP. Nutrition in the post-genomics: Fundamentals and applications of omics tools. Journal of Nutrition. 2008;**21**(6):757-766

[12] Dalmiel L, Vargas T, Molina AR. Nutritional genomics for the characterization of the effect of bioactive molecules in lipidmetabolismand related pathways. Electrophoresis. 2012;**33**(15):2266-2289

[13] Cozzolino SMF, Cominetti C. *Biochemical and Physiological Bases of Nutrition in Different Stages of Life in Health and Disease*. 1st ed. Monole Paulo, Brazil; 2013

[14] Sanhueza J, Valenzuela A. Nutrigenomics: Revealing the molecular aspects of a personalized nutrition. Chilean Journal of Nutrition. 2012;**39**(1):1-12

[15] Liu B, Qian SB. Translational regulation in nutrigenomics. American Society for Nutrition. 2011;**2**:511-519

[16] Ong TP, Rogero MM. Nutrigenomics: Importance of nutrient-gene interaction for health promotion. Journal of the ABESO. 2009;**40**

[17] Norheim F, Gjelstad IM, Hjorth M, Vinknes KJ, Langleite TM, Holen T,

et al. Molecular nutrition research—The modern way of performing nutritional science. Nutrients. 2012;**4**(12):1898-1944

[18] Fenech M, El-Sohemy A, Cahill L, Ferguson LR, French TAC, Tai ES, et al. Nutrigenetics and nutrigenomics: Viewpoints on the current status and applications in nutrition research and practice. Lifestyle Genomics. 2011;**4**(2):69-89

[19] Rana S, Kumar S, Rathore N, Padwad Y, Bhushan S. Nutrigenomics and its impact on life style associated metabolic diseases. Current Genomics. 2016;**17**(3):261-278

[20] Franzago M, Santurbano D, Vitacolonna E, Stuppia L. Genes and diet in the prevention of chronic diseases in future generations. International Journal of Molecular Sciences. 2020;**21**(7):2633-2667

[21] Afman L, Muller M. Nutrigenomics: From molecular nutrition to prevention of disease. Journal of the American Dietetic Association. 2006;**106**(4):569-576

[22] Phillips CM. Nutrigenetics and metabolic disease: Current status and implications for personalised nutrition. Switzerland. Nutrients. 2013;**5**(1):32-57

[23] Garg R, Sharma N, Jain SK. Nutrigenomics and nutrigenetics: Concepts and applications in nutrition research and practice. Acta Medica International. 2014;*1*(2):124

[24] Cozzolino SMF. *The Bioavailability of Nutrients*. 4th ed. Monole Paulo, Brazil; 2012

[25] Ordovas JM, Kaput J, Corella D. Nutrition in the genomics era: Cardiovascular disease risk and the Mediterranean diet. Molecular Nutrition & Food Research. 2007;**51**(10):1293-1299

[26] Corella D, Ordovas JM. Nutrigenomics in cardiovascular medicine. Circulation. Cardiovascular Genetics. 2009;**2**(6):637-651

[27] He P, Zhou R, Hu G, Liu Z, Jin Y, Yang G, et al. Curcumin-induced histone acetylation inhibition improves stressinduced gastric ulcer disease in rats. Molecular Medicine Reports. 2015;**3**:1911-1916

[28] Guimaraes MR, Leite FR, Spolidorio LC, Kirkwood KL, Rossa C. Curcumin abrogates LPS-induced pro-inflammatory cytokines in RAW 264.7 macrophages. Evidence for novel mechanisms involving SOCS-1, −3 and p38 MAPK. Archives of Oral Biology. 2013;**58**:1309-1317

[29] Santos AM, Lopes T, Oleastro M, Gato IV, Floch P, Benejat L, et al. Curcumin inhibits gastric inflammation induced by helicobacter pylori infection in a mouse model. Nutrients. 2015;7:306-320

[30] Gu B, Miao J, Fa Y, Lu J, Zou S. Retinoic acid attenuates lipopolysaccharide-induced inflammatory responses by suppressing TLR4/NF-kappaB expression in rat mammary tissue. International Immunopharmacology. 2010;**10**:799-805

[31] Singh U, Devaraj S. Vitamin E: Inflammation and atherosclerosis. Vitamins and Hormones. 2007;**76**:519-549

[32] Narasimhulu CA, Selvarajan K, Litvinov D, Parthasarathy S. Anti-atherosclerotic and anti-inflammatory actions of sesame oil. Journal of Medicinal Food. 2015;**18**:11-20

[33] Selhub J, Byun A, Liu Z, Mason JB, Bronson RT, Crott JW. Dietary vitamin B6 intake modulates colonic inflammation in the IL10−/− model of inflammatory bowel disease.

The Journal of Nutritional Biochemistry. 2013;**24**:2138-2143

[34] Kang R, Kim M, Chae JS, Lee SH, Lee JH. Consumption of whole grains and legumes modulates the genetic effect of the APOA5-1131C variant on changes in triglyceride and apolipoprotein A–V concentrations in patients with impaired fasting glucose or newly diagnosed type 2 diabetes. Trials. 2014;**15**:100

[35] Li S, Hu B, Wang Y, Wu D, Jin L, Wang X. Influences of APOA5 variants on plasma triglyceride levels in Uyghur population. PLoS One. 2014;**9**(10):110-258

[36] Hosseini-Esfahani F, Mirmiran P, Daneshpour MS, Mehrabi Y, Hedayati M, Zarkesh MM, et al. Western dietary pattern interaction with APOC3 polymorphism in the risk of metabolic syndromes: Tehran lipid and glucose study. Journal of Nutrigenetics and Nutrigenomics. 2014;**7**:105-117

[37] Lin J, Yang R, Tarr PT, Wu PH, Handschin C, Li S, et al. Hyperlipidemic effects of dietary saturated fats mediated through PGC-1beta coactivation of SREBP. Cell. 2005;**20**:261-273

[38] Gentschew L, Bishop KS, Han DY, Morgan AR, Fraser AG, Lam WJ, et al. Selenium, selenoprotein genes and Crohn's disease in a case–control population from Auckland, New Zealand. Nutrients. 2012;**4**:1247-1259

[39] Haritunians T, Jones MR, McGovern DP, Shih DQ, Barrett RJ, Derkowski C, et al. Variants in ZNF365 isoform D are associated with Crohn's disease. Gut. 2011;**60**:1060-1067

[40] Ferguson L. Nutrigenetics, nutrigenomics and inflammatory bowel diseases. Expert Review of Clinical Immunology. 2013;**9**:717-726

[41] Wang TT, Dabbas B, Laperriere D, Bitton AJ, Soualhine H, Tavera-Mendoza LE, et al. Direct and indirect induction by 1,25-di-hydroxyvitamin D3 of the NOD2/CARD15-defensin beta2 innate immune pathway defective in Crohn disease. The Journal of Biological Chemistry. 2010;**285**(4):2227-2231

[42] Ferretti G, Bacchetti T, Masciangelo S, Saturni L. Celiac disease, inflammation and oxidative damage: A nutrigenetic approach. Nutrients. 2013;**4**:243-257

[43] Romanos J, Van Diemen CC, Nolte IM. Analysis of HLA and non-HLA alleles can identify individuals at high risk for celiac disease. Gastroenterology. 2009;**137**:834-840

[44] Li S. Physical activity attenuates the genetic predisposition to obesity in 20,000 men and women from EPIC-Norfolk prospective population study. PLoS Medicine. 2010;**7**:1-9

[45] Boschiero C. Association of IGF1 and KDM5A polymorphisms with performance, fatness and carcass traits in chickens. Journal of Applied Genetics. 2013;**54**:103-112

[46] Ordovas JM, Kaput J, Corella D. Nutrition in the genomics era: Cardiovascular disease risk and the Mediterranean diet. Molecular Nutrition & Food Research. 2007;**51**(10):1293-1299

[47] Engbersen AMT, Franken DG, Boers GHJ, Stevens EMB, Trijbels FJM, Blom HJ. Thermolabile 5-10-methylenetetrahydrofolate reductase as a cause of mild hyperhomocysteinemia. American Journal of Human Genetics. 1995;**56**:142-150

[48] Fradet V, Cheng I, Casey G, Witte JS. Dietary omega-3 fatty acids, cyclooxgenase-2 genetic variation,

and aggressive prostate cancer risk. Clinical Cancer Research. 2009;**15**(7): 2559-2566

[49] Frayling TM, Timpson NJ, Weedon MN, Zeggini E, Freathy RM. A common variant in the FTO gene is associated with body mass index and predisposes to childhood and adult obesity. Science. 2009;**316**:889-894

[50] Dauncey MJ. Overview of nutrigenomics and epigenomics: Mechanisms and relevance. In: Proceedings of FACTA Avian Nutrigenomics Course, Campinas SP, Brazil; 2014

[51] Ferguson LR. Nutrigenomics approaches to functional foods. Journal of the American Dietetic Association. 2009;**109**(3):452-458

[52] Bae KC, Park JH, Na AY, Kim SJ, Ahn S, Kim SP, et al. Effect of green tea extract/poly-_-glutamic acid complex in obese type 2 diabetic mice. Diabetes and Metabolism Journal. 2013;**37**:196-206

[53] Mulvihill EE, Huff MW. Antiatherogenic properties of flavonoids: Implications for cardiovascular health. The Canadian Journal of Cardiology. 2010;**26**:17A-21A

[54] Nicolle E, Souard F, Faure P, Boumendjel A. Flavonoids as promising lead compounds in type 2 diabetes mellitus: Molecules of interest and structure-activity relationship. Current Medicinal Chemistry. 2011;**18**:2661-2672

[55] Chae YJ, Kim CH, Ha TS, Hescheler J, Ahn HY, Sachinidis A. Epigallocatechin-3-O-gallate inhibits the angiotensin II induced adhesion molecule expression in human umbilical vein endothelial cell via inhibition of MAPK pathways. Cellular Physiology and Biochemistry. 2007;**20**:859-866

Chapter 4

Value-Added Foods: Characteristic, Benefits, and Physical Properties

Zuzana Hlaváčová, Eva Ivanišová, Peter Hlaváč, Ľubomír Kubík, Daniela Kunecová, Monika Božiková and Vlasta Vozárová

Abstract

The growth of diet-related diseases (obesity, diabetes, osteoporosis, and cardiovascular diseases) is becoming an important societal concern and a challenge for a more sustainable society. This has developed important trends in food consumption, including the increasing demand for innovative food with natural attributes and with health claims (foods with added value, enriched foods, and functional foods). The physical properties of food are essential for scientists and engineers at solving the problems in food preservation, processing, storage, marketing, consumption, and even after consumption. In this chapter, we introduce the utilization of physical properties in enriched foods evaluation. The correlations were found between physical properties and other characteristics of foods.

Keywords: value-added food, health benefits, mechanical, rheological, thermophysical, and electrical properties

1. Introduction

Nowadays, people think deeply about how can their diet influence health and prefer to eat foods that can help them to get a better lifestyle. This situation support development of important trends in food consumption, which has seen, among others, the growing consumer interest in foods with natural and health properties. The category of food with health benefits includes enriched food with bioactive compounds, such as phenolics, mineral compounds, vitamins, and natural colorants, while healthy food is food without synthetic additives and human interventions, considered by consumers harmful for their organism [1]. The main goal of functional foods is to support the human body, strengthen immunity, or reduce the body's susceptibility to civilization diseases. This concept of functional foods is originated from Japan (1984) [2]. The Japanese government defined a new product category, Food for Specific Health Uses (FOSHU), as "food containing an ingredient with functions for health and officially approved to claim their physiological effects on the human body" and produced a dedicated legislative framework. Japan was followed by the United States, which in the '90s developed the first health claim act, but without providing a formal definition of functional food. European countries acquired the functional food concept more than 10 years later when the European Parliament and Council introduced the regulation on nutritional and health claims (Reg. (EU) n. 1924/2006); but also, in this case no formal definition was mentioned. The research interest in functional food experienced a steep increase only in the 21st century and this globally growing attention has tremendously influenced their market, the size of which was estimated at USD 162 billion in 2018 and was projected to reach USD 280 billion by 2025 with an annual growth rate of around 8% [3]. On the other hand, the term "nutraceuticals" is a combination of two words—nutrition and pharmaceutical; it is a substance that can be considered as food, which provides health benefits in that it can prevent or cure diseases. Nutraceuticals are health-promoting compounds or products that have been isolated or purified from food, providing a positive effect against some chronic diseases [4] and are generally sold in medical form (pills). Functional food products can be consumed as part of a daily diet, and medical food is sold under the supervision of doctors.

In recent years, value-added produce has gained increasing interest among small-scale produce growers. Agriculture defines one approach to adding value to a raw agricultural commodity as changing its physical state or form by processing and transforming it into a sellable product with a higher value. Some examples include making fruit jams from fruits or making fermented or pickled vegetables from vegetables. Previous studies also reported that growers were interested in adding value to their products by making jams and jellies. Value-added production enables growers to utilize excess fresh produce to make finished products to capture losses from unmarketable produce, provide more customer choice, and increase profitability [4, 5].

1.1 Plant-derived foods with added value

Plant-derived foods with added value are divided into primary and secondary metabolites; primary metabolites are plant components important for growth, while secondary metabolites are not essential for growth, but are important for plant survival mechanisms. Primary metabolites include proteins, beta-glucans, and omega-3 fatty acids. Plant proteins include texturized vegetable protein, soy protein isolate, and amino acids; these proteins act as functional ingredients in foods that can help to decrease the amount of meat consumption, connecting with positive effects to decreasing fat and cholesterol in the diet. Beta-glucans acting as foods with added value due to decreasing cholesterol absorption can be especially found in higher amounts in oat and barley. Omega-3 fatty acids, found in higher amounts in flaxseed and chia seed also act as a functional food due to their properties, such as reducing platelet aggregation. Phytoestrogens, antioxidants, vitamins, tocopherols, steroids, and gamma-linolenic acid are considered as other very important secondary metabolites. Phytoestrogens, compounds in plants, especially found in soybeans, flaxseed, and alfalfa plants, are used more to produce enriched foods due to the possibility of decreased post-menopausal cancer development. Antioxidants, such as flavonoids, act as functional compounds by quenching reactive oxygen species. Vitamins act as functional compounds by preventing deficiencies and are rich in fruits and vegetables; certain vitamins, such as vitamins C and E, also act as quenchers of reactive oxygen species. Tocopherols, which belong to vitamin E compounds found in oilseeds, act as quenchers of reactive oxygen species. In oilseeds, steroids are found that act as functional components and reduce the absorption of cholesterol [5, 6].

Agricultural and agri-food systems are influenced by trends that could jeopardize their future sustainability. The population is growing very quickly and people have started getting afraid about their food sources and are changing their dietary preferences. Persistent poverty, inequality, and unemployment constrain access to food and hamper the achievement of food security and nutrition goals. Agricultural production is limited by the increasing scarcity and diminishing quality of land and water resources, as well as by insufficient investment in sustainable agriculture. Climate change is increasing, which has a negative impact, affecting yields and rural livelihoods, while agriculture continues to emit large amounts of greenhouse gases [7]. The global increase in demand for food and the limited land area available prompt the search for alternative food sources, rich in biologically active compounds. Medicinal herbs, edible flowers, and wastes from food industries (coffee skin, and cacao shell) can be interesting for the production of foods with added value. Edible flowers have become a culinary trend, referred to in international culinary magazines, such as Bon appetite "How to use edible flowers in salads, cocktails, and more" and Food and Wine "The Edible Flower." The clients of edible flowers are gourmet restaurants and their associated food service operations, and grocery stores. So, eating edible flowers is a new trend, described as one of the "six trends of food and drinks in gastronomy" [8, 9]. Edible flowers possess nutritional value—being rich in moisture, carbohydrates, and protein, and being low in lipids. They also contain interesting amounts of ash, including dietary minerals, such as calcium, iron, potassium, magnesium, phosphorous, and zinc. Furthermore, they contain bioactive components, such as phenolic compounds, which contribute to their high antioxidant activity, while also conferring color and aroma. Other biological effects include antimicrobial and anti-inflammatory activities that are also reported to inhibit cell proliferation, turning them into a potential ally for cancer treatment and prevention. Still, it is important to bear in mind what amounts need to be ingested for these health effects to be effective on the human body. From this point of view, many of these possible health claims are not yet established through recommended intake dosages [9, 10].

1.2 Animal-derived foods with added value

Animal-derived foods with added value are enriched mainly with bioactive compounds, such as omega-3 and six fatty acids, conjugated linolenic acid, small peptides, whey and casein, and glucosamine and chondroitin sulfate. Omega-3 fatty acids include alpha-linolenic, docosahexaenoic, and eicosapentaenoic fatty acids. Soy and canola oils, walnuts, and flaxseed belong to one of the best sources of alpha-linolenic acid. The main animal source of these fatty acids is fatty fish, such as salmon, tuna, and cod. Omega-3 and six fatty acids have a positive impact on immunity, modulating inflammation, and protecting against neurodegenerative diseases. Omega-6 fatty acids include linolenic, gamma-linolenic, and arachidonic fatty acids. Vegetable oils, nuts, and whole grains are considered as main plant sources. Conjugated linoleic acid is a fatty acid present in milk and related products that reportedly acts as a functional compound due to its properties to reduce cancer risks and adipose differentiation; however, a fatty liver may develop as a side effect. Whey and casein are milk proteins that act as functional ingredients by being easily digested and absorbed, and help in building muscle mass; small peptides function in the same manner. For collagen formation are required glucosamine and chondroitin sulfate, which as functional compounds, can alleviate pain associated with osteoarthritis; however, this claim must be supported by more observations [5]. Attractive source for the production of animal-derived foods in the food industry can be edible insects.

Edible insects as an alternative protein source for human food are interesting in terms of low greenhouse gas emissions, high feed conversion efficiency, low land use, and their ability to transform low-value organic side streams into high-value protein products. More than 2000 insect species are eaten mainly in tropical regions [11]. Eating insects by humans is not a new concept; it occurs globally but is still rare in Europe. Why not eat insects? Is it worth it? The answer is simple—definitely yes. Entomophagy has several advantages. First of all, insects are a good source of protein, essential fats, and antioxidant peptides. Many insects are rich in microelements, such as iron, calcium, zinc, and vitamins. Secondly, insect breeding is environmentally friendly. Insects emit significantly fewer greenhouse gases and ammonia than most livestock. Moreover, insects require less space, feed, and water for breeding than livestock. Economic factors are also important. Insect rearing can be low-tech or very sophisticated, depending on the level of investment. For these three main reasons, insects have been highlighted as an important food source in response to the growing concerns about the future of world food security [12, 13].

1.3 Microbial-derived foods with added value

Value-added foods derived from microbial sources include probiotics, prebiotics, symbiotics, and synbiotics. Probiotics are natural microflora that occurs in the gut, such as *Lactobacillus casei* or numerous *Bifidobacterium* species, which improve health. Prebiotics are dietary compounds that promote the growth of probiotic bacteria. Symbiotics contain both—probiotics and prebiotics, combined randomly, while synbiotics contain specific probiotics and prebiotics mixed together to benefit one another. Value-added foods derived from microbial sources promote the growth of probiotic bacteria so that the growth of pathogenic bacteria is limited and these properties can improve the immune system of the human body [5].

1.4 Miscellaneous-derived foods with added value

Some sources and foods with added values are derived from miscellaneous compounds, such as algae and mushrooms [5]. Nowadays global demand for foods enriched by macroalgal and microalgal components is growing, and algae are increasingly being consumed for health benefits beyond the traditional considerations of nutrition and health [14]. Algae contain a large number of bioactive components with several positive biological activities, such as antioxidant, antimicrobial, and anticarcinogenic. Algae are also considered as a source of fiber in the form of sulfated galactans or carragenates in red algae, fucans, alginates, and laminarin. A diet rich in dietary fiber has a positive impact on human health that reduces the risks of cancer, cardiovascular diseases, diabetes, obesity, hypercholesterolemia, and digestive problems. The diet with higher dietary fiber content showed good immunological activity. Alginate in *Undaria pinnatifida* showed a positive effect on cardiovascular disease, and alginic acid was demonstrated to reduce hypertension in hypertensive rats. Alginic acid, xyloglucans in *Sargassum vulgare* and sulfated fucans in *Undaria pinnatifida* have shown powerful anti-viral activity against herpes type-1 and cytomegalovirus in humans, which can be used in future medicine and pharmacy [15].

The rising demand for functional food free from synthetic chemicals indicates the awareness of people on quality food. The excellent texture and unique flavor of edible and medicinal mushrooms make them universally accepted by all age groups. Due to the production of a large variety of secondary metabolites with exceptional chemical structures and interesting biological actions they are a reservoir of valuable chemical resources. However, there is very little awareness of mushrooms as healthy food and as an important source of biologically active substances with medicinal value [16]. Mushrooms are highly nutritive, low-calorie food with good quality proteins, vitamins, and minerals. Mushrooms are an important natural source of foods and medicines. By virtue of having high fiber, low fat, and low starch, edible and medicinal mushrooms have been considered to be an ideal food for obese persons and for diabetics to prevent hyperglycemia. They are also known to possess promising antioxidative, cardiovascular, hypercholesterolemia, antimicrobial, hepatoprotective, and anticancer effects. More than 3000 mushrooms are mainly edible species but, only 100 species are cultivated commercially, and only 10 species are used at industrial scale and their global and economic value is now increasing slowly due to an increase in their value as a food as well as their medicinal and nutritional values [17].

1.5 Perspectives of foods with added value

Cereal products (bread, pastries, pasta, cookies, bars, etc.) are an important source of macro and micronutrients that are essential for human health and are widely consumed by all classes of society. Their high consumption, as a snack has earned it the matrix for food fortification. Research has been done on the enrichment of these products with legumes, oil crops, nuts, medicinal herbs as well as new alternative food sources (edible insects, flowers, algae, and mushrooms). Studies have shown that enrichment can influence the physicochemical and nutritional properties of cereals products. The regular consumption of enriched cereal foods can reduce the risk of several forms of cancer, coronary heart diseases and helps to regulate blood glucose levels and cholesterol, chemical and nutritional properties of cereals products.

Figure 1.
Algae, edible flowers, and medicinal mushrooms as a perspective source for innovative cereal-based products with added value (Ivanišová).

Mechanisms of the effect of phytochemicals from alternative food sources—edible insects, flowers, algae, and medicinal mushrooms are the subject of many planned studies with a link to practical use. These perspective sources can be attractive for enriched innovative cereal-based products with added values (**Figure 1**).

2. Mechanical properties of candies

The quality of the confectionery represents the main aim of scientific studies. Authors [18] are interested in the study of the influence of process operating conditions and candy composition on the unsteady behavior of the cooling process of hard candies to improve final product quality. Hard candy is often sold as drops, lozenges, barrels, rods, and sticks. Hard candy comes in the form of ribbons and disks of cut rock, with colorful and unique shapes worked into the interior of each piece. Commercial hard candies, with water content in the range from 2–4%, typically have glass transition temperature values about 25 to 45°, depending on the sugar composition and moisture [19]. The sensory analysis is often applied to the determination of the firmness of the candies. The firmness can be slightly affected by the presence of watermelon flours that conferred a certain fibrous and graininess to the candy, even if most of the sensory scores were in the category of like very much (score 8) [20].

The aim of the study is the determination of the mechanical properties of the honey candies by compression method and discussion of the firmness and brittleness of the material during storage in dependence on the variation of the water content.

2.1 Material and methods

The mechanical properties of eight sorts of honey candies prepared by a selected company were studied by methods of compression between two parallel metal plates. Four sorts of honey candies were made from cane sugar and the next four sorts of honey candies were made from coconut sugar. Each sort of honey candies was coated in a different sort of starch or powdered sugar. There were four possibilities, coating with the powdered sugar (saccharose), corn starch, rice starch, and potato starch. We obtained eight combinations of the samples coated with different sugar. Samples of candies were then stored for 2 months at room temperature in paper wrappers and a test measurement was realized every 30 days. There were performed three measurements, at the beginning of the storage, after 30 days, and after 60 days. Five samples of each sort of coated candies were measured by the testing compression equipment for each of eight combinations of the candies coating.

Experimental compression equipment was developed at the Institute of electrical engineering, automation, informatics, and physics of the Slovak University of Agriculture in Nitra, Slovak Republic. The equipment includes the stepped motor that serves on the control of the screw shifting, power source, screw, PLC modules, and PLC software. Equipment was composed of the PLC X20CP3583 with processor Intel Atom. The universal mixing external module X20DM9324 with eight digital inputs and four digital outputs was used with a module of bus X20BM11 and terminal board X20T12. Digital servo amplifier Acopos 8V1016 was used for the control of the stepped motor.

The force sensor LT Lutron FG 6100SD was used for the measurement of the loading force. The sensor worked up to maximal load 1000 N with uncertainty ±0.1 N. Sensor of displacement WDS-500-P60-SR-U (Micro-Epsilon, Czech Republic) with

14-bit A/D converter, up to 500 mm with uncertainty ±0.5 mm and with the software uncertainty ±0.001 mm, was used.

2.2 Results of the compress properties

The firmness of the confectionery can determine as the force needed for the compression between teeth or between tongue and palate, from the sensoric point of view. Force (N) and compression (mm) of the candies were measured by test compression equipment. Dependencies of the force on the compression were determined by the compression of the candies between two parallel metal plates of the compression equipment. The measurement was performed at the speed of compression of 10 mm.min^{-1}. Failure strength was obtained from the maximum of the loading curves as the index of the confectionery quality. The compression diagrams of coconut sugar coated with powdered sugar for the five samples immediately after manufacturing are presented in **Figure 2** as an example of the 40 measurements, which were realized.

The regression equations of the curves from **Figure 2** are shown in **Table 1**. Failure strength was obtained from the peaks of the curves. The peaks represent the firmness of the candies. The samples were measured immediately after manufacturing and they had low moisture, therefore they were brittle, and they cracked at the low loading up to 300 N. Crack was observed on the samples with the peaks. The sample without a peak was not cracked. Failure strength of the coconut candies coated with the powdered sugar was in the level 105–355 N and the compressions were about 1.4 mm.

The mechanical properties of all combination of coconut coating candies are presented in **Table 2**.

The compression diagrams of cane sugar coated with powdered sugar for the five samples immediately after manufacturing are presented in **Figure 3** as an example of the 40 measurements, which were realized. The regression equations of the curves

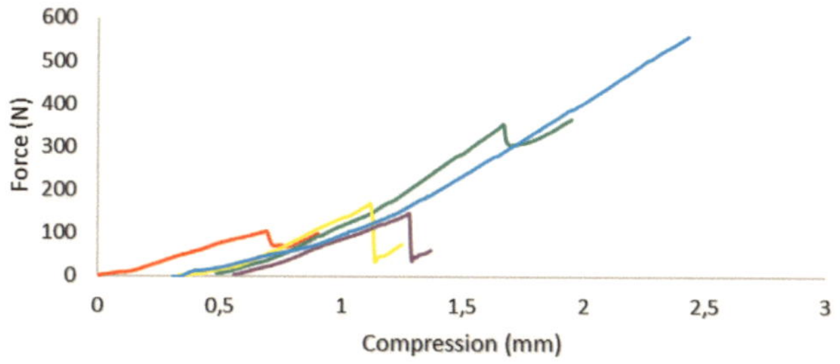

Figure 2.
Compression diagram of five honey candy samples with coconut sugar coated with powdered sugar.

Candies	Coating	n	Regression equations	Coefficient of determination
Coconut sugar	Powdered sugar	1	$y = 33.4x^2 + 139.24x - 4.5261$	$R^2 = 0.9941$
		2	$y = 121.85x^2 + 37.017x - 45.107$	$R^2 = 0.9995$
		3	$y = 158.54x^2–1.5876x - 2.,776$	$R^2 = 0.9954$
		4	$y = 39.189x^2 + 130.99x - 84.432$	$R^2 = 0.985$
		5	$y = 104.34x^2–0.4932x - 4.6281$	$R^2 = 0.9991$

Table 1.
Regression equations of the honey candies with coconut sugar coated by with powdered sugar.

	Coconut sugar—powdered sugar		Coconut sugar—corn starch		Coconut sugar—rice starch		Coconut sugar—potato starch	
n	Force	C	Force	C	Force	C	Force	C
	(N)	(mm)	(N)	(mm)	(N)	(mm)	(N)	(mm)
1	103.6	0.690	128.9	0.659	207.2	1.530	205.6	1.110
2	349.7	1.659	201.4	1.029	368.9	2.009	120.2	1.019
3	166.0	1.110	558.8	2.029	111.5	0.840	508.1	1.620
4	147.2	1.279	352.2	1.579	322.8	1.179	529.0	1.770
5	540.6	2.379	387.7	1.319	356.4	1.439	248.1	1.269
Average	261.413	1.423	325.783	1.323	273.356	1.399	322.194	1.357
SD	182.187	0.637	168.122	0.522	110.752	0.433	185.193	0.325
VC (%)	69.693	44.787	51.605	39.480	40.515	30.978	57.478	23.945

C, compression; SD, standard deviation; VC, coefficient of variation.

Table 2.
Mechanical properties of honey candies made from coconut sugar after the first measurement at the beginning of storage.

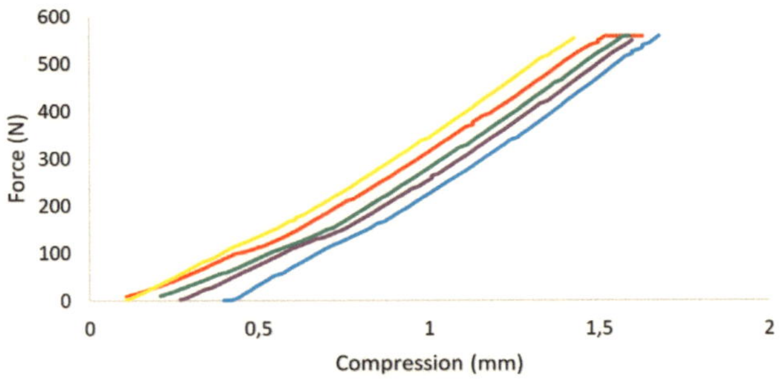

Figure 3.
Compression diagram of five honey candy samples with cane sugar coated with powdered sugar.

from **Figure 3** are shown in **Table 3**. Dependencies are realized without peaks. The candies did not crack at the compression up to 500 N. The compression reached values of about 1.6 mm. The candies were brittle but simultaneously harder than coconut sugar candies.

Candies	Coating	n	Regression equations	Coefficient of determination
Cane sugar	Powdered sugar	1	$y = 74.653x^2 + 270.29x - 34.46$	$R^2 = 0.9964$
		2	$y = 118.84x^2 + 198.26x - 41.114$	$R^2 = 0.9993$
		3	$y = 93.38x^2 + 278.75x - 28.937$	$R^2 = 0.9997$
		4	$y = 113.05x^2 + 202.94x - 59.155$	$R^2 = 0.9996$
		5	$y = 90.033x^2 + 255.38x - 118.71$	$R^2 = 0.9997$

Table 3.
Regression equations of the honey candies with cane sugar coated with powdered sugar.

Measurement of the honey candies realized at the beginning of storage was characterized by the cracking of the coconut candies of all coatings at the low forces. Cane candies did not crack even if had low moisture. The phenomena could be the cause of the bigger hardness. The structure and the size of the starch molecules affect the mechanical properties of the starch coating. The amount of amylose influence the firmness of the starch coatings and gels and the molecular profile of amylopectin participates in the properties of the films [21].

The mechanical properties of all combination of cane coating candies are presented in **Table 4**.

2.2.1 Results after 30 days of storage

Mechanical properties of the honey candies were measured after 30 days of storage at room temperature in paper wrappers. Again, were realized measurements of the coconut and cane candies of the coating by the powdered sugar (saccharose), corn starch, rice starch, and potato starch. There were realized 40 compressions of all combinations of the candies. Failure strength was obtained from the peaks of the curves. The candies had bigger moisture and there was more plastic because the samples did not crack. The mechanical properties of all combination of coconut coating candies are presented in **Table 5**. The mechanical properties of all combination of cane coating candies are presented in **Table 6**.

The samples after 30 days of storage were not so brittle but were more plastic, accordingly, the bigger forces were needed for deformation. The partial sticking occurred, but the samples were able to separate. Cane candies obtained after 30 days of storage had higher failure strength than at the beginning of storage. The cracking of the samples was observed at a force of 600 N. The coconut candies were more brittle than cane candies. The moisture of the candies was augmented in the direction from the border to the center of the confectionery.

	Cane sugar—powdered sugar		**Cane sugar—corn starch**		**Cane sugar—rice starch**		**Cane sugar—potato starch**	
n	**Force**	**C**	**Force**	**C**	**Force**	**C**	**Force**	**C**
	(N)	**(mm)**	**(N)**	**(mm)**	**(N)**	**(mm)**	**(N)**	**(mm)**
1	556.5	1.620	524.8	1.370	533.2	1.579	530.0	1.620
2	556.7	1.589	534.2	1.579	514.2	1.659	557.8	1.800
3	542.4	1.409	510.2	1.719	526.0	1.569	521.0	1.569
4	547.4	1.599	497.3	1.740	528.9	1.679	537.3	1.610
5	557.2	1.679	507.1	1.279	520.3	1.630	557.9	1.780
Average	552.016	1.579	514.711	1.537	524.522	1.623	540.793	1.676
SD	6.737	0.101	14.654	0.206	7.419	0.048	16.616	0.106
VC (%)	1.221	6.417	2.847	13.430	1.414	2.976	3.072	6.339

C, compression; SD, standard deviation; VC, coefficient of variation.

Table 4.
Mechanical properties of honey candies made from cane sugar after the first measurement on the beginning of storage.

	Coconut sugar—powdered sugar		Coconut sugar—corn starch		Coconut sugar—rice starch		Coconut sugar—potato starch	
n	Force	C	Force	C	Force	C	Force	C
	(N)	(mm)	(N)	(mm)	(N)	(mm)	(N)	(mm)
1	806.8	2.270	830.6	2.346	821.2	3.9	819.6	2.750
2	285.0	1.840	710.0	5.426	703.4	4.0	671.4	1.640
3	779.2	3.020	490.6	0.680	426.8	3.2	804.4	9.170
4	530.0	3.230	622.2	2.140	809.8	2.6	814.2	3.080
5	218.4	3.030	815.4	3.020	832.6	3.0	723.4	7.190
Average	523.880	2.678	693.760	2.722	718.760	3.324	766.600	4.766
SD	243.167	0.532	126.633	1.552	153.113	0.521	59.047	2.896
VC (%)	46.417	19.860	18.253	57.024	21.302	15.665	7.702	60.775

C, compression; SD, standard deviation; VC, coefficient of variation.

Table 5.
Mechanical properties of honey candies with coconut sugar after 3 days of storage.

	Cane sugar—powdered sugar		Cane sugar—corn starch		Cane sugar—rice starch		Cane sugar—potato starch	
n	Force	C	Force	C	Force	C	Force	C
	(N)	(mm)	(N)	(mm)	(N)	(mm)	(N)	(mm)
1	703.4	2.030	837.6	2.770	811.4	2.430	837.8	1.217
2	816.8	2.040	801.2	1.220	805.6	2.380	810.0	2.260
3	802.6	2.340	656.8	2.280	804.2	2.200	634.2	1.640
4	802.0	1.740	418.8	1.640	644.0	1.320	826.2	2.540
5	851.2	2.860	798.2	2.850	834.2	3.150	827.2	2.160
Average	795.200	2.202	702.520	2.152	779.880	2.296	787.080	1.9634
SD	49.256	0.380	154.743	0.635	68.790	0.586	76.955	0.473
VC (%)	6.194	17.248	22.027	29.509	8.821	25.516	9.777	24.114

C, compression; SD, standard deviation; VC, coefficient of variation.

Table 6.
Mechanical properties of honey candies with cane sugar measurement after 30 days of storage.

2.2.2 Results after 60 days of storage

Mechanical properties of the honey candies measured after 60 days of the storage at room temperature in paper wrappers were realized also as 40 compressions of all combinations of the candies. The candies had a high amount of moisture at the time and they were plastic. The moisture content of the candies was from 4.5 to 5.5%, which was the reason for the gluing of the candies. It was not possible to separate them. It was not possible to measure coconut candies. The cracking was not observed. The candies had a texture like krowka candy. Only cane candies were measured. The

	Cane sugar—powdered sugar		Cane sugar—corn starch		Cane sugar—rice starch		Cane sugar—potato starch	
n	Force	C	Force	C	Force	C	Force	C
	(N)	(mm)	(N)	(mm)	(N)	(mm)	(N)	(mm)
1	828.2	2.540	816.8	1.570	804.0	1.200	825.000	2.14
2	804.2	1.540	807.2	2.310	852.4	1.730	829.000	1.93
3	825.8	1.970	821.6	1.900	810.0	2.890	716.000	2.13
4	852.0	2.430	816.0	1.670	805.4	1.550	803.200	2.27
5	814.4	2.170	845.0	3.350	814.4	2.380	215.200	1.88
Average	824.920	2.130	821.320	2.160	817.240	1.950	677.680	2.07
SD	17.922	0.398	14.221	0.724	20.076	0.678	262.541	0.161
VC (%)	2.173	18.682	1.731	33.500	2.456	34.775	38.741	7.797

C, compression; SD, standard deviation; VC, coefficient of variation.

Table 7.
Mechanical properties of honey candies with cane sugar after 60 days of storage.

mechanical properties of all combination of cane coating candies are presented in **Table 7**. The moisture of the candies significantly increased after 60 days of storage. The gluing of the cane candies was smaller than the coconut candies. The moisture of the cane candies was at the level of 4.5–5.5%. The cane glued candies were possible to separate and measure. The coconut glued candies had moisture at the level of 7–6%. The glued coconut candies changed their shape and the measurement was impossible. The brittleness of the hard candies was decreased in consequence of the increased moisture. The amount of moisture content of the hard confectionery is from 0.8 to 1.2%. Hard candies with a bigger amount of moisture than 2.5% are displayed by the deformations without cracks [22].

The confectionaries were only compressed, but not cracked after compression. Hard confectionaries made from cane sugar save relatively high hardness, even if condensed at relatively high temperatures.

Results obtained at the measurement of the firmness and hardness of the confectionaries demonstrate that the important parameter indicating the firmness or brittleness of the material is the measure of the saturation of the material by the wet air. Different proportion of moisture affects the plasticity of the hard candies. An increasing of the moisture inflicts a decrease of the hardness and the increasing of the plasticity of the material. Increased moisture during storage contributes to the gluing of the hard candies.

3. Rheologic properties

Rheological parameters could be used in many areas, such as product quality evaluation, engineering calculations, and process design. Understanding the flow behavior is important, for example, in the determination of the size of the pump and pipe and energy requirements. For designing food engineering processes in connection with energy and mass balances, rheological models from experimental measurements can be used. Influence of processing on rheological properties must be known for process control [23, 24].

Fluids could be divided by Newton's law of viscosity

$$\tau = \eta \text{ grad } v \tag{1}$$

where τ is shear stress; grad v is the gradient of velocity (in many cases called as shear rate).

Newtonian fluids are following this law. Dynamic viscosityη, as the slope in the shear stress shear rate dependency, is constant and independent of the shear rate for these liquids [25].

The viscosity of Newtonian materials is influenced only by temperature and material composition, and it is independent of the shear rate and previous shear history. Examples of Newtonian materials are oils, water, beer, wine, milk, clear fruit juices, sucrose solutions, most honey, cream (till fat content of 40%), butter (above the melting point), etc. [24, 26, 27].

For the characterization of Newtonian fluids is also often used kinematic viscosityν, which is defined as a ratio between dynamic viscosity and densityρ of fluid at the same temperature

$$\nu = \frac{\eta}{\rho} \tag{2}$$

The physical unit of kinematic viscosity is $m^2 \cdot s^{-1}$, but for liquids is more often used unit $mm^2 \cdot s^{-1}$. An important role is also played by the fluidity φ of materials, which can be expressed as a reciprocal value of dynamic viscosity and that is why its unit is $Pa^{-1} \cdot s^{-1}$ [28].

$$\varphi = \frac{1}{\eta} \tag{3}$$

Values of fluids' dynamic viscosity are changing from 10^{-5} Pa·s for some gases to values of the order of Pa·s for some oils.

Oppositely there are non-Newtonian fluids that are affected by the shear rate. For this type of liquid, apparent viscosity η is applied, which is expressed in the same units as dynamic viscosity (Pa·s or in mPa·s) [26].

On the base of the shearing time, non-Newtonian foods could be separated into two groups—time-independent and time dependent. Apparent viscosity of time-independent fluids is affected only by the shear rate, but in the case of time-dependent fluids, it is also the time duration of the shearing. In time-independent flow, shear-thinning (pseudoplastic) and shear-thickening (dilatant) behavior could be observed, which depends upon whether the viscosity decreases or increases with an increase in shear rate [26]. Shear-thinning and shear-thickening fluids could be described by the power law model (Ostwald—de Waele equation)

$$\tau = K \, (\text{grad } v)^n \tag{4}$$

where K is coefficient of consistence and n is flow behavior index [25]. For the shear-thinning (pseudoplastic) fluids is index n lower than 1 and for the shear-thickening fluids is this index higher than 1. Newtonian liquids can be considered as a special case of this model, where n is 1 and K corresponds to dynamic viscosity. The

DOI: http://dx.doi.org/10.5772/intechopen.104971

slope in stress—rate dependency is not constant for non-Newtonian fluids. In pseudoplastic flow behavior, it could be observed that shear stress is increasing while viscosity (or friction between layers) is decreasing with increasing shear rate, therefore it is called shear-thinning behavior [24, 25]. Due to the shearing, entangled, long-chain molecules are straightening out and become aligned with the flow, therefore it reduces the viscosity. Different shear rates mean different viscosities and for non-Newtonian fluids apparent viscosity η_a is applied, which can be defined as a ratio of shear stress and corresponding shear rate

$$\eta_a = \frac{\tau}{grad\ v} \tag{5}$$

When we apply Eq. (4) together with previous Eq. (5) then apparent viscosity could be derived

$$\eta_a = K\ (grad\ v)^{n-1} \tag{6}$$

Into shear-thinning materials could be included concentrated fruit juices, melted chocolate, French mustard, mayonnaise, dough, butter, honey, dairy cream, fruit and vegetable purees, etc. [23, 24, 26, 29].

Fluids, for which increasing in shear rate results in an increase in internal friction and apparent viscosity, are called shear-thickening fluids. When the increase in viscosity is connected with volume expansion, these fluids are called dilatant fluids. Dilatant flow behavior could be observed when the shear stress is increasing with the shear rate increase. Viscosity is also increasing with increasing shear rate and that is why this case is referred to as shear-thickening behavior [24, 25].

Shear-thickening materials are not very common among the food materials, but this behavior can be observed in some honey, corn starch suspension, etc. [26, 24].

Bingham plastic fluids are fluids that remain rigid when the value of shear stress is smaller than the yield stress τ_o, but when it overcomes this value they flow like a Newtonian fluid. That is why these liquids could be used in this equation

$$\tau = \tau_0 + K\, grad\ v \tag{7}$$

For example, mayonnaise, tomato paste, ketchup, concentrated pulpy and granular suspensions, chocolate can be included into Bingham plastic fluids [29, 24].

When the stress is lower than the yield stress then the material behaves like a solid, its energy is stored in small strains and a flat surface is not obtained under the influence of gravity. This behavior is necessary for designing processes and assessing the quality of materials, such as butter, yogurt, and cheese spreads [29]. Sometimes also in non-Bingham plastic fluids, yield stress must be applied and only after it exceeds its value, the flow is enabled. In this case, the dependency of shear stress on the shear rate is not linear like in Bingham plastic fluids [24, 25].

There are other models for describing the non-Newtonian fluid behavior, for example, Herschel—Bulkley model valid for minced fish paste, raisin paste, and rice flour-based batter, Casson model which was used for molten milk chocolate [23–25].

Non-Newtonian foods with time-dependent flow properties can be subdivided into thixotropic and rheopectic fluids. In thixotropic fluids, at a fixed shear rate, the viscosity decreases with time, whereas the viscosity of a rheopectic fluid increases with time [26].

Fluids for which shear stress and viscosity are decreasing with time at a fixed shear rate are called thixotropic fluids (shear-thinning fluids with time). Shear thinning with time is caused by the breakdown of the material structure during the continuous shearing. The reason for decreasing viscosity is assumed to be caused by a decrease in the intermolecular interactions within the molecular structure of the material [25, 24]. Thixotropic behavior has been observed for various materials, such as gelatine, condensed milk, mayonnaise, butter, honey, dough, and egg white. [26, 23, 24].

Another type of fluid is rheopectic fluid (shear-thickening fluids with time), where shear stress and viscosity increase with time at a fixed shear rate, and the material structure builds up during the continuous shearing. Viscosity (or shear rate) increase arises from the intermolecular interactions that cause the friction to increase with time at a constant shear rate within the molecular structure of the material [25, 24]. Rheopectic behavior is also referred to as antithixotropic behavior. Antithixotropic behavior in food systems is very rare [23].

Fluid foods go through different processes, such as processing, storage transportation, marketing, and consumption, where the temperature is changing. That is why the rheological properties are examined in the temperature dependencies. Arrhenius equation could be used for expressing the influence of temperature on the viscosity, which is set at a certain shear rate [26].

For overcoming energy barriers thus enabling the flow of the fluid food material, the energy is needed, and it is called activation energy E_A. This energy is more readable at higher temperatures and the flowing of the fluid is easier. The dependence of viscosity on the temperature could be modeled by an Arrhenius type equation

$$\eta = \eta_0 e^{-\frac{E_A}{R\,T}} \tag{8}$$

where η_0 is the initial value of dynamic viscosity; R is gas constant; and T is absolute temperature [25, 24].

3.1 Rheologic properties utilization

The piezoelectric-excited membrane device was used for rapid measurement of liquids viscosity and density, the principle is based on the membrane's resonant frequency and Q factor responses to the damping effects of a surrounding liquid. The authors performed measurements on five samples of oils, which had viscosity in the interval (19.88–1733) mPa·s and density in the range (829–886) kg·m^{-3}. Authors stated that viscosity and density measurement could be used at monitoring the oil quality and the determination of the pipeline elements' design; the texture of liquid foods could be achieved by maintaining the viscosity and density control during the production, which will cause high production efficiency and cost-effectiveness; for the treatment of certain vascular diseases, the blood fluidity control is required, so the blood viscosity must be monitored [30]. The density of food materials is needed in many areas, for example in separation processes, pneumatic and hydraulic transports, determination of the power required for pumping, etc. [23].

The food spoilage extent by using an activatable molecular rotor was also examined. The authors measured the viscosities of various liquids (white spirit, pure milk, green tea, edible vinegar, fresh broth, and lemon juice). For fresh broth and lemon juice, the authors examined the influence of different storing conditions on the viscosity values. The viscosity of the fresh broth and lemon juice increased more rapidly at an ambient temperature of 25°C within 9 days and on the other hand, a slower increase was observed in samples stored at a fresh-keeping temperature of 4°C [31].

Texture properties of yogurts (with wheat bran; with different amounts of date fiber; with date fiber and vanilla flavor) were compared [32]. The author's study has shown that fortifying yogurt with 3% of date fiber is an acceptable product with potential beneficial health effects.

Rheological behavior of Galician's honey was also examined [33]. The authors performed measurements on 11 samples of honey. According to the dependency of apparent viscosity on the shear rate, the authors had concluded that measured honey can be classified as non-Newtonian fluids but indicated that most honey can be included in Newtonian fluids. Non-Newtonian rheological behavior has been detected when low values of shear rate have been applied, whereas, for high values of shear rate, honey tends to have a Newtonian fluid behavior. The authors observed pseudoplastic behavior and fitted it with the Ostwald model. They also proved that honey with higher water content has the lowest viscosity values, whereas honey with high sugar contents shows elevated values of apparent viscosity [33].

The dynamic oscillatory shear rheological characteristics of honey samples (pine, citrus, and flower) from different floral sources were evaluated at three different temperatures (10°C, 15°C, and 20°C). The authors found that all honey samples showed liquid-like behavior (Newtonian flow behavior) [34]. The shear stress of Turkish honey samples increased, while the apparent viscosity remained constant with increasing shear rate. The authors obtained highest complex viscosity for pine honey at 10°C, while the lowest value was found for citrus honey at the same temperature. Temperature dependency of complex viscosity of honey samples was modeled by the Arrhenius model [34].

Rheologic properties of natural and reduced-calorie Israeli honey were also analyzed. Authors found that the viscosity of honey was Newtonian, even in reduced-calorie varieties, and adhered to the Arrhenius equation, where viscosity exponentially decreases with temperature [35].

The temperature and time of constant temperature heating on the rheological properties of light (apple, Ziziphus, and citrus) and dark (black horehound, globe thistle, and squill) types of honey were examined. The authors in their research found that both honey types behave like a Newtonian fluid regardless of the conditions of heating. Another finding of the authors was that light-colored honey, which had lower water contents than dark-colored ones, after heat treatment showed a change in viscosity only at higher heating temperatures in comparison with the fresh untreated control sample. On the other hand, dark-colored, heat-treated honey showed a change in viscosity at all levels of heating temperature [36].

The influence of temperature on the rheological behavior of Portuguese honey was also investigated. The authors found that all honey showed flow independence over time and behaved as Newtonian fluids at the studied temperature and shear rate ranges. For all honey, the authors observed that the viscosity decreased with temperature and good regression coefficients were obtained by fitting the experimental data with the Arrhenius model [37].

At the determination of the temperature dependency of viscosity was observed that viscosity of creamed honey decreased with increase in temperature level for which they applied the Arrhenius equation. The authors had explained that when the thermal expansion occurs as a result of an increase in temperature, less energy is required to breakdown the structure at higher temperature levels due to the reduction of intermolecular forces, increasing intermolecular distance between the molecules, thus causing the reduction of intermolecular forces, finally the decrease in viscosity [38].

The jams typically display a shear-thinning behavior associated with a tendency to reach a yield stress value. This flow behavior has been frequently expressed in terms of the Herschel-Bulkley model. Authors [39] found that viscosity values increased with pectin concentration and decreased with increasing temperature. Arrhenius-type relationships were used to quantify the influence of temperature on the storage modulus, yield stress, and consistency index. The authors also found higher temperature dependence of linear viscoelastic functions for model borojó jam formulations in comparison with the peach jam sample, but it was still lower than borojó commercial jam. On the contrary, the influence of temperature on the yield stress was lower for both commercial jams, whereas the effect on the consistency index was very similar for all examined jam samples [39].

The structural parameters (particle content, particle size, and serum viscosity) influence the rheological properties of Golden Delicious apple purees. It was observed that all the purees presented a shear-thinning behavior, yield stress, and they were not time dependant. The apparent viscosity, yield stress, and elastic modulus decreased as particle size decreased and they increased as insoluble solids content increased. Herschel Bulkley's model was used for the estimation of the yield stress [40].

The flow behavior of two commercial date pastes was of interest. The shear stress–shear rate data were fitted by using six famous models known as Power-law, Bingham plastic, Herschel–Bulkley, Casson, Sisko, and Vocadlo. It was observed that date pastes exhibited the shear-thinning pseudoplastic behavior and that the Casson model was best for describing the experimental data at all temperatures. For the temperature dependence of apparent viscosity, authors used the Arrhenius model with a very high coefficient of determination [41].

Influence of composition on vegetable oils oxidation using differential scanning calorimetry was studied under non-isothermal conditions at five different heating rates, in a temperature range of 100–400 °C [42]. Eight vegetable oils were examined—refined palm oil, olive oil, grapeseed oil, sunflower oil, corn oil, soybean oil, safflower oil, and sesame oil. Activation energy of measured oils was also determined, and the authors found that refined palm oil exhibited the highest values of activation energy. Authors declared that vegetable oils play an important role in the human diet due to their nutritional and sensory properties. Authors also pointed out that oils are susceptible to oxidation reactions during food processing and subsequent storage of food products [42].

The rheological behavior of the pure BIO chocolate at different temperatures was also studied. Authors had observed that the chocolate unambiguously demonstrated plastic behavior and flow curves were fitted by the power-law model (Herschel–Bulkley model), Bingham model, and Casson model, with the best coefficient of determination for the Casson model, but the Herschel-Bulkley model also gave very accurate results of pure chocolate flow curve. It also examined the temperature dependence of apparent viscosity of chocolate in the range from 35°C up to 62°C. It was found that the apparent viscosity decreases in the studied temperature range and this decrease could be fitted by using the power law equation. Authors found from flow curves of pure chocolate that temperature increase had caused shear stress decrease [34].

Changes in tomato powder during storage at different storage temperatures were evaluated for 5 months. Higher storing temperatures had a significant effect on tomato powder qualities, while lower temperature had less effect. Tomato powders have many advantages, including ease of packing, transportation, and mixing. In addition, tomato powder can be used as an ingredient in many food products, mainly soups, sauces, and ketchup. The consumption of tomato and tomato-based products has been associated with a lower risk of developing certain types of cancers [44].

The nutritional, rheological, and sensory properties of tomato ketchup with increased content of natural fibers made from fresh tomato pomace were evaluated, and the results were compared with five commercial products. The rheological properties of the ketchup with increased fiber content depend mostly on total solids and insoluble particle content, but properties remained within the limits for standard tomato products. Rheological properties of tomato products are considered as one of the most important quality attributes since they influence product processing parameters, especially flow properties during transport, as well as consumers' acceptability [45].

It was stated that ultrasonic pulses could be combined with rheological measurements at the determination of solids concentration in different highly concentrated and industrial food suspensions, such as tomato, vegetable and pasta sauces, seafood powder, strawberry yogurt, and cheese sauce with vegetables. Low-intensity ultrasound-based techniques could be applied to the detection and identification of foreign bodies in food products [46].

3.2 Measurement results

In the following part, examples of rheologic properties measurements are presented. Majority of analyzed materials were Newtonian materials and therefore these properties were studied—dynamic viscosity, kinematic viscosity, and fluidity. In the case of non-Newtonian material, apparent viscosity was determined. Effect of various factors on these properties was investigated. Relations of dynamic (or apparent), and kinematic viscosity to the temperature can be described by decreasing exponential functions Eqs. (9) and (10), and in the case of relations of fluidity to the temperature can be used increasing exponential function Eq. (11).

$$\eta = C\, e^{-D\left(\frac{t}{t_0}\right)} \tag{9}$$

$$\nu = E\, e^{-F\left(\frac{t}{t_0}\right)} \tag{10}$$

$$\varphi = G\, e^{H\left(\frac{t}{t_0}\right)} \tag{11}$$

where t *is* temperature; t_0 = 1°C, C, D, E, F, G, H—constants dependent on the kind of material, and on ways of processing and storing. A similar model was used by the author [47].

Temperature dependencies of dynamic viscosity for milk different fat content are presented in **Figure 4**. From **Figure 4** it is clear that the highest dynamic viscosity values were observed for the sample milk with the highest fat content (3.5%) and the lowest dynamic viscosity values were obtained for the sample milk with the lowest fat content (0.5%). The highest viscosity value 1.99 mPa·s was obtained at the lowest measured temperature for milk with the highest fat content and oppositely the lowest dynamic viscosity 1.30 mPa·s was obtained at the highest temperature for milk with the lowest fat content. Decreasing exponential function can be used for the description of presented dependencies and that is in accordance with Arrhenius Equation [23, 25, 43, 48, 49].

Figure 5 depicts temperature dependencies of kinematic viscosity observed in analyzed samples of vegetable oils. For both samples, decreasing function describes the dependencies of kinematic viscosity on temperature. In contrast to extra virgin olive oil, sunflower oil showed higher values of kinematic viscosity. This is most likely due to different compositions of oils.

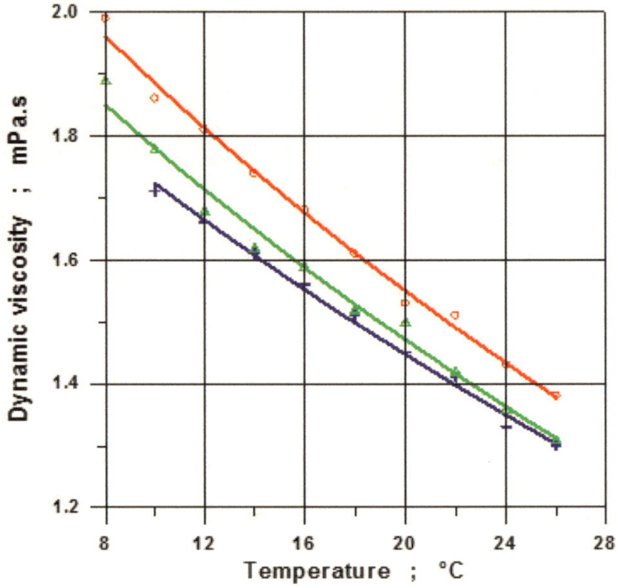

Figure 4.
Temperature dependencies of milk dynamic viscosity: Fat content 0.5% (+); fat content 1.5% (△); fat content 3.5% (○).

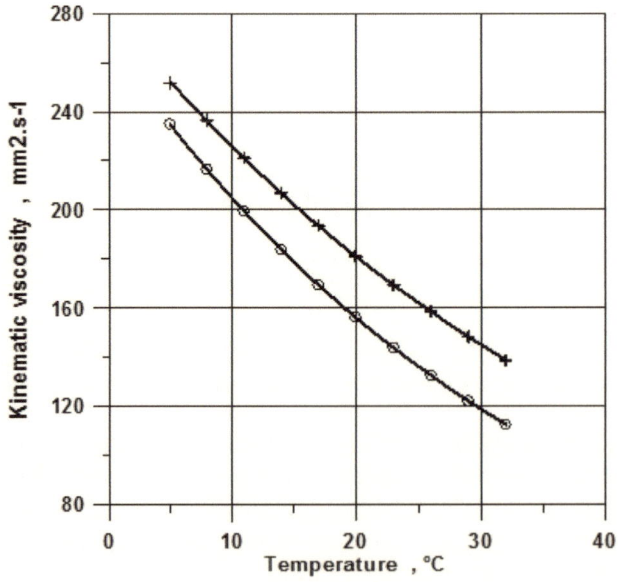

Figure 5.
Temperature dependencies of vegetable oils kinematic viscosity, sunflower oil (○), extra virgin olive oil (+).

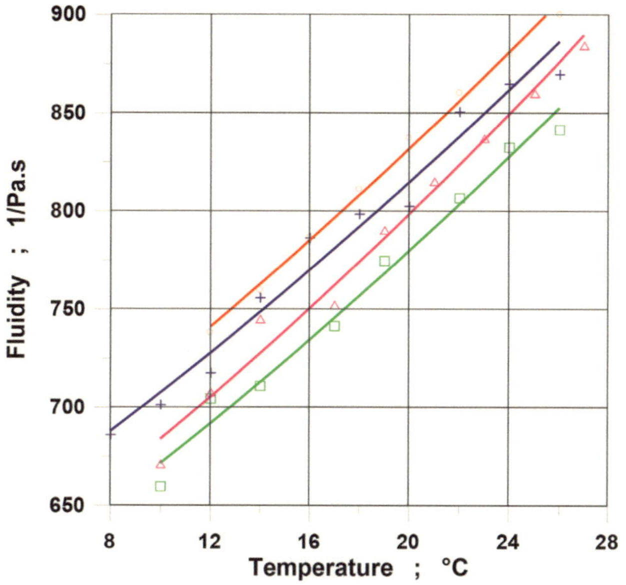

Figure 6.
Temperature dependencies of beer fluidity: Topvar 10% (+), Topvar 12% (□), Zlatý Bažant 10% (○), Zlatý Bažant 12% (△).

Fluidity increases with temperature for all beer samples (**Figure 6**). The proportion of the curves could be caused by the alcohol content because the highest alcohol content 5.4% had beer sample Topvar 12%, and it had the lowest fluidity, and on the contrary the lowest alcohol content 4.3% had beer sample Zlatý Bažant 10%, which had the highest fluidity.

In **Figure 7**, temperature dependencies of ketchup's apparent viscosity are presented. It is possible to observe that the apparent viscosity of the ketchup is decreasing with increasing temperature. The progress can be described by decreasing exponential function, which is in accordance with Arrhenius Eq. (8).

Comparable rheological results for ketchup were reported [50]. Approximately similar values of the parameter in the regression equation contributing to apparent viscosity

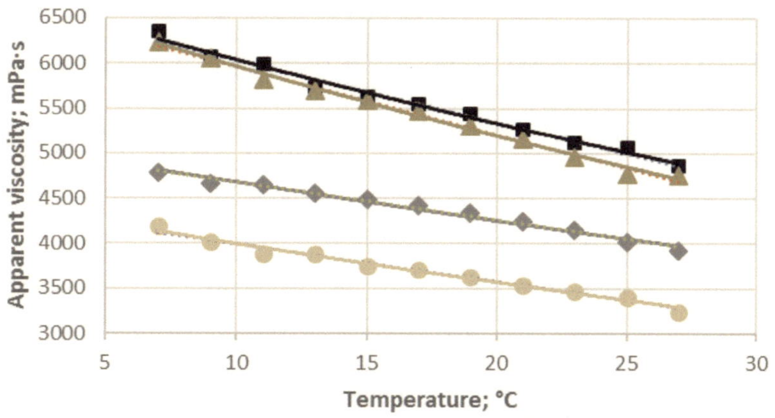

Figure 7.
Temperature dependencies of ketchup apparent viscosity–Snico (■), Hellmanns (▲), Heinz (◆), Hamé (●).

can be found in the paper [51]. It is also visible in **Figure 7** that the highest apparent viscosities were obtained for a sample of Ketchup Snico, and the lowest values for Ketchup Hamé, which could be caused by different compositions of ketchup.

It could be observed that density and rheologic properties are influenced by several factors, including fat content, alcohol content, time of storing, composition, concentration, and even added ingredients.

4. Thermophysical properties

4.1 Calorimetric methods

The thermal analysis includes techniques in which a physical property of a substance is measured as a function of temperature [52]. The chosen calorimetric analysis depends on the type of research. Different calorimetric methods can be used for quantitative and qualitative evaluation of biomass, polymeric materials, and various types of foods produced industrially or even fresh fruits and vegetables.

A better understanding of the temperature influence on the food properties allows food manufacturers to improve the processing and reforming quality of products. It is, therefore, important for food science to have analytical techniques to monitor the temperature changes that occur in foods. These techniques are often grouped under the general heading of thermal analysis. Most foods are subjected to variations in their temperature during production, transport, storage, preparation, and consumption. It is possible to use thermal analysis for tracking pasteurization, sterilization, evaporation, cooking, and freezing.

4.2 Thermogravimetric analysis (TGA)

The main applications of TGA are processes, such as evaporation, desorption and vaporization behavior, thermal stability, kinetics of decomposition, and compositional analysis. Thermogravimetric techniques connect the measurement of sample mass during the heating or cooling at a controlled rate (heating rate) or at a set isothermal part for a time. Thermogravimetry is useful for monitoring processes that involve a change in the mass of a food or its individual parts. Thermogravimetric instruments have been specially designed to allow measurements to be carried out under specific environments, controlled pressures, or atmospheres. The mass of a sample may change with temperature or time depending on the occurring specific physicochemical processes. Heating often leads to a reduction in mass because of the evaporation of water or other volatile parts. On the other hand, the mass of food may increase due to the moisture absorption from the atmosphere at specific conditions of storage. The ability to carefully control the temperature, pressure, and composition of the gasses surrounding a sample is extremely valuable for food sciences because it could be used for modeling processes, such as drying, cooking, and absorption of moisture during the storage. Determination of moisture and water distribution in food could be done by thermogravimetric analysis. It is possible to set the final drying temperature as well as the heating rate and a certain temperature to end the drying [53, 54]. The drying process is recorded at each performed analysis. In the process of heating to temperatures above 300°C, it is possible to diagnose the quantity and quality of food composition. The TGA plot is an overlay of thermograms of the same type of food from different manufactured batches showing variations. It demonstrates that TGA is a valuable tool to check if materials conform to set decomposition requirements. The result of TGA measurements is usually displayed as a TGA curve (thermogram), temperature or time dependence of mass. Another important information is the use of the first derivative of the TGA curve with respect to temperature or time (usually linear dependence). This shows the rate at which the mass changes and is known as the differential thermogravimetric (DTG) curve.

Mass changes occur when the sample loses material in one of several different ways or reacts with the surrounding atmosphere. This leads to the production of steps in the TGA curve or peaks in the DTG curve. A number of different effects can cause loss, or even gain of mass in the sample and produce steps in the TGA curves. TGA thermal decomposition thermogram of biological materials generally follows similar patterns in 2 to 4 steps. Foods are multi-ingredient materials therefore more steps are expected [55]. Another important factor in TGA or other thermal analysis is the heating rate. The systematic deviation exists between the true sample temperature and the measured temperature, and it is heating-rate dependent. Real samples (mixtures) can of course exhibit quite different thermal conductivity behavior. If the sample undergoes chemical reactions, the temperature region in which the reaction occurs is very much dependent on the heating rate. In general, higher heating rate causes reactions to shift to higher temperatures. If unsuitable heating rates are used, the reactions may overlap and remain undetected. A quite different approach for separating overlapping reactions is based on the use of DTG.

4.2.1 Measurement results

Measured materials in the thermogravimetric analysis were gluten-free biscuits with dried and crushed additions of agrimony (*Agrimonia eupatoria* L.), littleleaf linden (*Tilia parvifolia* L.), breckland thyme (*Thymus serpyllum* L.), and oregano (*Origanum vulgare* L.), respectively. TG analysis was performed in the nitrogen atmosphere from 25–700°C by a heating rate of 10°C.min^{-1}. The analyzer TGA/DSC 1 from Mettler Toledo (Switzerland) was used at measurement. Flow of carrying gas was 50 ml.min^{-1}. Biscuits were weighted in alumina crucibles with lids. Whole volume of the crucible is 70 µl, it has a cylindrical shape and the circle lid is pierced. Mass of all materials was from 13.9 mg to 22.8 mg. On the thermogram (**Figure 8**) is visible 4 decreases that are common to all samples. First decline corresponds with the evaporation of water in biscuits.

These steps are most affected by storage and are observed at 150°C. Around 250°C, it is the decomposition of sucrose; the average decrease corresponds to 9.6% of mass loss. The second and third decreases from 300–450°C correspond to the decomposition of the starch structures of the biscuits. These are the largest components of the food, as the average decrease is from 19–30% for the temperature range (300–380) °C and 16–30% for temperatures above 390°C. Percentage decrease corresponds to mass loss and the amount of the components in samples. The second, third, and fourth mass loss are similar and therefore no major change in the quantity of composition is expected. Obvious change in decline is in steps of evaporation, where biscuits with agrimony and Breckland thyme had the greatest decline.

4.3 Differential scanning calorimetry (DSC)

A differential scanning calorimeter measures the heat flow that occurs in a sample when it is heated, cooled, or held isothermally (constant temperature). The technique

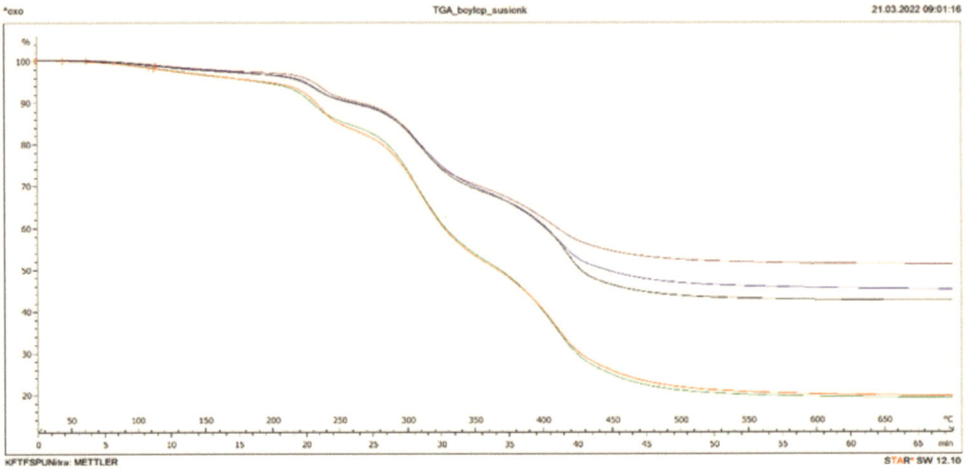

Figure 8.
TGA of gluten-free biscuits (— Agrimony; — Breckland thyme; — Control sample; — Littleleaf linden; — Oregano).

is also called differential scanning calorimetry (DSC). It allows to detect endothermic and exothermic effects, measure peak areas (reaction enthalpies), determine temperatures that characterize the peak or other effects, and specific heat capacity c. Since this amount of supplied heat causes a corresponding increase in the enthalpy (H) of the sample, we can write Eq. (12) (thermodynamic definition).

$$C_p = \frac{dH}{dT} \tag{12}$$

where C_p is heat capacity at constant pressure; $\frac{dH}{dT}$ is the heat capacity and thus the slope of the enthalpy—temperature function.

The most important effects that can be analyzed by DSC are the melting point, melting range, and melting behavior. DSC is used to determine the heat of fusion, purity, polymorphism, glass transition, and oxidation stability [56]. Food researchers and food industry have shown an increased interest in techniques that can predict modifications in quality and thermophysical properties of food products during processing and storage. Differential scanning calorimetry (DSC) has attracted the interest of food scientists because only a small amount of sample is needed for analysis and to give exact results [57]. The DSC thermal method is used to describe and determine thermal processes in food. It is a differential scanning calorimetry that can be used to determine the process temperature and temperature requirement for an ongoing process in food. It is possible to determine the purity and authenticity of oils. For ingredient foods, it is possible to determine the decomposition temperature of the food ingredient and to describe the phase events in the food [58, 59]. The DSC method can be implemented alone or as an associated DTA to TGA method. DSC is rapid, facile, and capable of supplying both thermodynamic (heat capacity, enthalpy, and entropy) and kinetic data (reaction rate and activation energy) on protein denaturation. For independent implementation of DSC, it is possible to examine food even at sub-zero temperatures and to monitor events, such as crystallization or solidification of individual food components. At higher temperatures, it is possible to monitor the breakdown of proteins or the melting of individual mixtures of food ingredients. Differential scanning

DOI: http://dx.doi.org/10.5772/intechopen.104971

calorimetry can be used to characterize mixtures of polymorphic forms of fats as well as to evaluate moisture and various thermal load for their effectiveness in bringing about desired polymorphic changes. DSC has also been employed for the investigation of the physical state and water content in foodstuffs [60].

In special DSC device settings, it is also possible to determine specific heat capacity, where the specific heat capacity for DSC is determined from Eq. (13).

$$Q = mc_p \frac{dT}{dt} \tag{13}$$

where Q *is* heat flow per time unit; m *is* sample mass; c_p is specific heat capacity; and $\frac{dT}{dt}$ *is the* rate of external temperature change.

The specific heat capacity (c) is significant information in the food industry and it means the amount of energy that must be supplied or taken from material to change its temperature by a given amount [58]. The specific heat capacity of a material determination is therefore important in the design of processes, such as storage at low or high temperature and treatment of foods.

Dynamic DSC measurements are usually plotted as dependencies on temperature rather than on time. If the reference temperature is chosen as the main axis, the curve remains linear with time and does not change in appearance. DSC curves are distorted if they are plotted with respect to sample temperature.

4.3.1 Measurement results

The tested samples were value-added biscuits with the addition of dried and crushed orange, cole, and raspberry. Temperature was elected for the determination of melting of cocoa butter. According to the literature, cocoa butter melts in the temperature range (32–36°C) [59, 61].

DSC measurement was performed in the nitrogen atmosphere from 0–70°C by heating rate of 10°C.min^{-1}. The analyzer DSC 1 from Mettler Toledo (Switzerland) was used at measurement. Flow of carrying gas was 50 ml.min^{-1}. Biscuits were weighted on aluminum crucibles with lids. Whole volume of crucible is 100 μl.

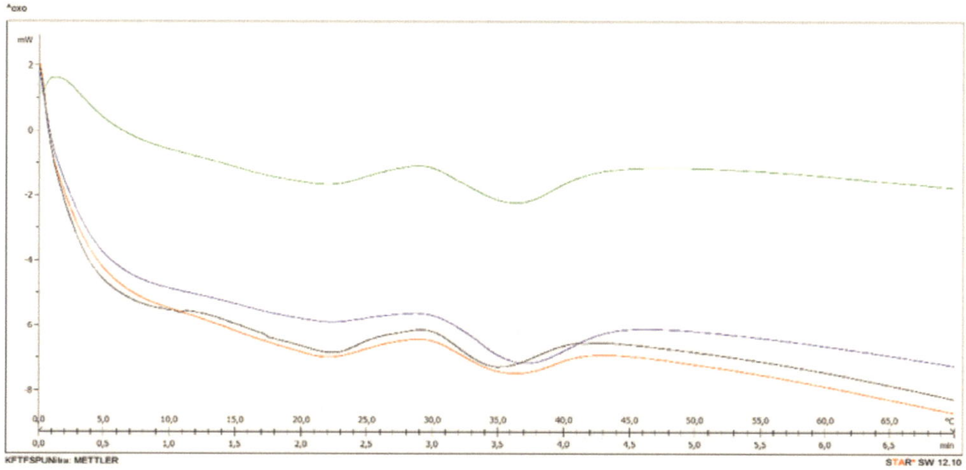

Figure 9.
DSC curves of biscuits (— Raspberry; — Cole; — Only cole powder, — Orange).

Sample	Normalized enthalpy (J·g^{-1})	Onset temperature (°C)	Peak temperature (°C)
Cole powder biscuit	2.48	30.11	34.47
Cole biscuit	3.70	29.93	36.11
Orange biscuit	4.60	30.60	36.45
Raspberry biscuit	2.50	30.06	35.56

Table 8.
Results of biscuit DSC measurement.

Mass of all materials was from 13.6 mg to 14.4 mg. On the DSC (**Figure 9**) is visible one peak, which is common to all samples. An endothermic peak at a temperature of about 36°C can be seen for all biscuit samples with various compositions.

This peak represents the melting of the cocoa butter contained in the biscuits. According to the shape and size of the peak, this is not a significant component of the investigated food. The start of the process can be set at 29.93°C. The addition of milk fat to the chocolate lowers its melting point, and palms fat can increase the melting point of the chocolate. However, a large temperature difference in the melting process is not seen, indicating a minimum content of fats in the sample. Total heat that is needed for the process of melting cocoa butter as is show in **Table 8** is from 2.5 J·g^{-1} to 4.6 J·g^{-1}.

5. Electrical properties

Electrical properties are utilized in many areas of human activities. They have the biggest application in moisture content measurements. The research on moist material is very important. The wide spectrum of electromagnetic waves can be used for material quality control and moisture content determination in many industries, for example, woodworking, civil engineering, also in agriculture, commerce, food quality investigation, and so on (e.g., [62]).

The moisture content is a very important property of materials. It is determined by various types of material, for example, soil, fruits, vegetables, cereals, foods, timber, and so on. Moisture content profiles in the material are also of interest. The distribution of moisture affects physical and biological processes, such as drying; moisture and heat transport, solute movement; uniformity of dye absorption in textiles; and the quality of many food products (e.g., [63]). The application of a method for noninvasive, nondestructive moisture profile measurement is very wide [63].

Electrical properties characterize the transport of charge carriers in materials or the propagation of electromagnetic waves in materials. In the group of electrical properties can be included—conductance G (S), which is an ability to conduct the electrical current, and the conductivity σ (S.m^{-1}), it is the conductance in relation to the proportions of the material sample. Reciprocal value of conductance is the resistance R (Ω), and reciprocal value of conductivity is resistivityρ (Ω.m).

Electrical conduction of the material is influenced by the chemical composition of material. Various types exist in the materials. Electronic conductance is characteristic of metals, charge carriers are electrons, and it occurs in conductors. Ionic type of conductivity is characterized by the fact that the charge carriers are ions and it occurs mainly in electrolytes. This type of conductivity also applies to food materials if they have sufficient moisture. Another type—hole conductivity or P-type conductivity exists in semiconductors; charge carriers are positively charged particles. There is also electrophoretic conduction; charge carriers are macromolecules or particles aggregation, which can be used in biological materials. The material resistance in case of alternate current is impedance Z (Ω). Alternate conductance is admittance Y (S) (e.g., [63]).

When the electric current travels through the biological material the electric conductivity type changes. The ionic conductivity occurs inside the cell. In cellular membrane exists displacement current. The definition of the electric current density i is

$$i = \frac{1}{S}\frac{dQ}{d\tau} \tag{14}$$

where Q *is* charge (C); τ is time (s); S is surface (m^2) (e.g., [63]).

The current density is the function of electric field intensity

$$\vec{i} = \sigma\vec{E} = -\sigma grad\, U \tag{15}$$

where σ is electric conductivity ($S{\cdot}m^{-1}$); grad U is gradient of the electric voltage ($V{\cdot}m^{-1}$); E is intensity of the electric field ($V{\cdot}m^{-1}$).

This equation is also valid for electrolytes at low values of electric field intensity.

If the current passing through the material is unsteady, for density of electric current is valid

$$i = \sigma E + \frac{d\varepsilon E}{d\tau} \tag{16}$$

where ε is the permittivity of material ($F{\cdot}m^{-1}$).

Material permittivity characterizes the ability of matter to polarize (form dipoles). Vacuum has the lowest permittivity because it contains no particles and no polarization occurs. Its permittivity is denoted by ε_o and has a value of 8854.10^{-12} $F{\cdot}m^{-1}$, which is one of the basic physical constants. To find out how many times the permittivity of a given medium is greater than the permittivity of the vacuum, the relative permittivity ε_r (also called dielectric constant) is used, which is defined as the ratio of the permittivity of the medium and the permittivity of the vacuum (e.g., [64]).

The relative permittivity of food materials depends mainly on the presence of water. Water is a polar liquid and simply polarizes, it has a high relative permittivity of up to 81 at low frequencies. Relative permittivity of food mineral components is from 3 to 7; for organic material 2 to 5; for ice 3 (e.g., [28]).

In materials of biological origin, conductive parts (cell interiors) alternate with non-conductive parts (cell membranes). This alternation creates biological capacitors whose surface capacity reaches about 1 $\mu F.cm^{-2}$. There is only a negligible amount of free charge carriers in the cell membranes. When they are placed in an electric field, polarization occurs and dipoles form. Dipoles can move in the direction of an external electric field—this mfovement is called a displacement current.

If the material is situated in an alternating electric field, we have to use the complex value of current density in the shape

$$\hat{i} = (\sigma + j\omega\varepsilon)\hat{E} \tag{17}$$

where $\hat{E}$ is complex value of electric field intensity; j is imaginary unit (1); ω is angular frequency (s^{-1}).

In the case of moist material, permittivity is a complex value. It consists of a real part ε' and an imaginary part ε''

$$\hat{\varepsilon} = \varepsilon' - j\varepsilon'' = \varepsilon - j\frac{\sigma}{\omega} = \varepsilon(1 - tg\delta) \tag{18}$$

where δ is the loss angle (1) of the dielectric and

$$tg\delta = \frac{\sigma}{\omega\varepsilon} \tag{19}$$

where tgδ is tangent of loss angle and also loss factor (1) (e.g., [63]).

ε_r'' is interpreted to include the energy losses in the dielectric arising from all-dielectric relaxation mechanisms and ionic conduction. The imaginary part characterizes the dielectric losses occurring in the material and represents the ability of energy dissipation in the dielectric, by which the energy of the electric field is converted into thermal energy in the dielectric [62–64].

The penetration depth d_p is the depth at which the waves power decreases to 1/e (about 37%) of the surface power of the material. In materials where dielectric losses exist, it is possible to calculate this depth according to the relationship [65].

$$d_p = \frac{c}{2\pi f\sqrt{2\varepsilon'\left(\sqrt{1+\left(\frac{\varepsilon''}{\varepsilon'}\right)^2} - 1\right)}} \tag{20}$$

where c is light velocity in a vacuum; f is frequency.

5.1 Electrical properties utilization

The electrical impedance measurement was used in the detection of chemical substitutes in liquid food products [66]. Impedance spectroscopy was used [67] on tea quality control. In the paper [68], a method for quality control and detection of oil adulteration that used microwave electrical properties and Cole-Cole dielectric parameters was described. An open-ended coaxial sensor was developed and used [69] by Abbas et al. [69] to determine the oil and water content of olive trees as well as their ripeness. Other authors [70] determined the activation energy of fatty acids obtained from olives in the microwave region. Many authors have also dealt with milk properties [71, 72]. In another paper [73], the deactivation of Escherichia coli bacteria in milk at high pulsed electric field intensities was described. New method for determining the water content of milk using measurements of dielectric properties in the radio and microwave range was developed [74].

The electrical properties of fruits and vegetables have also been investigated in many works. Correlations between electrical properties and quality indicators of different types of fruits and vegetables were sought. The greatest growth has been recorded in non-destructive methods for evaluating or controlling fruit quality parameters, including internal errors, but also taste, sugar content, etc. [26, 75, 76]. Further studies have examined the usefulness of using the dielectric properties of apples to detect changes in their quality during storage [77]. The quality of the apple pulp during storage was assessed based on electrical impedance and Cole-Cole diagrams [78].

The electrical properties of bulk materials, especially grains and seeds used in the food industry, have been in the area of interest for many years. They were mainly used to determine their moisture content (e.g., [63, 64]), moisture content in chick-pea flour [65], and it was found [79] a relationship between dielectric properties and the oatmeal admixture in wheat flour.

5.2 Electrical properties measurement results

The samples of value-added foods were prepared by the Faculty of Biotechnology and Food Sciences (**Figure 10**).

The pasta control sample had the following recipe—wheat flour, salt, water, and egg. Other samples contained a 3% mixture of dried and crushed nettle (Urtica dioica L.), carrot (Daucus carota L.), or elderberry (Sambucus nigra L.), respectively. The highest values of conductivity were achieved by the pasta with elderberry, lower pasta with carrots, and the lowest by pasta with nettle and a control sample. Up to a frequency of 20 kHz, the conductivity of the samples is difficult to distinguish. At higher frequencies, it would be possible to use measurements of electrical properties to assess admixture in the basic type of control sample. The nettle and control pasta samples have relatively the same and low values compared to the others. The dry matter content of the individual pasta was almost the same. Thus, the displacement of dependencies was caused by the added materials.

We also measured the electrical properties of six samples of value-added chocolates. One was a control without an addition. Other samples included admixture as candied pulp of butternut pumpkin (*Cucurbita moschata* L.), dried capuchin (*Tropaeolum majus* L.) leaves, candied young spruce shoots (*Picea abies* L.), dried peppermint (*Mentha piperita* L.) leaves, candied josta fruits (*Ribes nidigrolaria* L.), and dried rosehip leaves. The preparation of the samples was as follows—cocoa beans from Madagascar first underwent fermentation, followed by drying, roasting, peeling, grinding, and conching, which lasts 6 hours. The first layer of chocolate is poured into a mold, allowed to partially solidify, a layer of added materials (thickness about 2 mm) is applied, then another layer of chocolate and the whole is allowed to solidify.

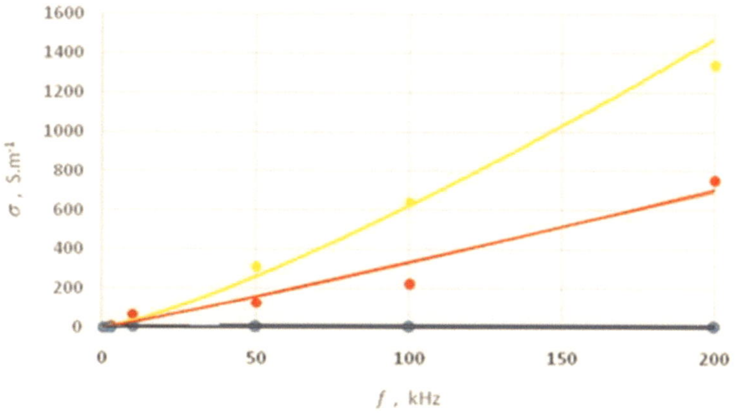

Figure 10.
Conductivity of pasta samples versus frequency (•—control, •—pasta with nettle, •—pasta with carrot, •—pasta with elderberry).

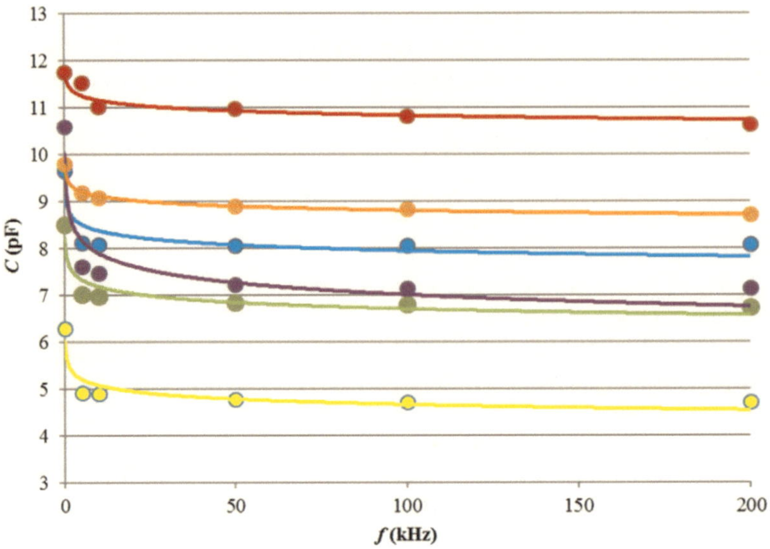

Figure 11.
Frequency dependencies of various chocolate capacity (●—Candied pulp of butternut pumpkin, ●—Control, ●—Dried capuchin leaves, ●—Candied young spruce shoots, ●—Dried peppermint leaves, ●—Candied josta fruits and dried rosehip leaves).

In **Figure 11**, we can see that the capacity decreases with increasing frequency for all samples. At all frequencies, the highest capacity values were measured on a sample of chocolate with a pumpkin. In contrast, the lowest values were measured on a sample with added josta and rose. Dependencies on **Figure 11** can be modeled using a regression function that has the form

$$C = C_0 \left(\frac{f}{f_0} \right)^{-k} \tag{21}$$

where C_0 is reference value of capacity; k is constant, f_0 = 1 kHz.

The regression equation has very high coefficients of determination for all samples of value-added chocolates. Similar dependencies were also obtained for relative permittivity. In this case, the displacement of curves is caused also by ingredients added to chocolate. It means that the electrical properties could be used in the recognition of material added to the chocolate.

The process of pasta and chocolate enriching was effective because it can increase antioxidant activity as well as the content of biologically active compounds.

The measurements of electrical properties in most cases have no effect on foods, it is a non-destructive measurement. Measuring with electrical sensors is quick, easy, and reliable and no operator is required, while the resulting data can be directly processed into the required electronic form. Dielectric spectroscopy techniques enable nondestructive and noninvasive measurements of the agricultural materials and foods, therefore providing tools for rapid evaluation of their water content and quality [65, 80, 81].

6. Conclusions

Innovative foods with added value, being one of the major food categories of the global health and wellness market, are becoming a major focus of new product development in the food industry. The development of these kinds of foods is more complex than traditional food, calling for a concerted effort from researchers and experts to explore and understand the process in more detail (technology, nutritional properties, hygienic, sensory, and physical properties).

Results obtained at the measurement of the firmness and hardness of the confectionaries demonstrate that the important parameter indicating the firmness or brittleness of the material is the measure of the saturation of the wet air by the material. Different proportion of moisture affects the plasticity of the hard candies. Increasing the moisture inflicts a decrease in the hardness and an increase in the plasticity of the material. Increased moisture during storage contributes to the gluing of the hard candies.

It could be observed that density and rheologic properties are influenced by several factors, including temperature, fat content, alcohol content, time of storing, composition, concentration, and even added ingredients.

Most foods are subjected to variations in their temperature during production, transport, storage, preparation, and consumption. It is possible to use thermal analysis for tracking pasteurization, sterilization, evaporation, cooking, and freezing.

Electrical properties measurement enables nondestructive and noninvasive measurements of the agricultural materials and foods, therefore providing tools for rapid evaluation of their quality and also distinguishing the added material. We found out that a correlation exists between electrical properties and the type of admixture added to the foods.

Further research in the area of physical properties is needed to clarify the impacts of value-added components on food quality.

Acknowledgements

This research has been co-funded by the European Community under the project No 26220220180: Building the Research Centre AgroBioTech and was supported by the Operational Program Integrated Infrastructure within the project: Demand-driven research for the sustainable and innovative food, Drive4SIFood 313011 V336, co-financed by the European Regional Development Fund.

Author details

Zuzana Hlaváčová*, Eva Ivanišová, Peter Hlaváč, Ľubomír Kubík, Daniela Kunecová, Monika Božiková and Vlasta Vozárová
Slovak University of Agriculture in Nitra, Nitra, Slovakia

*Address all correspondence to: zuzana.hlavacova@uniag.sk

References

[1] Tóth J, Migliore G, Schifani G, Rizzo G. Sustainable value creation in the food chain: A consumer perspective. Sustainability. 2020;**12**:1-12. DOI: 10.3390/su12041438

[2] Grochowicz J, Fabiashak A, Ekielski A. Importance of physical and functional properties of foods targeted to seniors. Journal of Future Foods. 2021;**1-2**:146-155. DOI: org/10.1016/j.jfutfo.2022.01.004

[3] Alongi M, Anese M. Re-thinking functional food development through a holistic approach. Journal of Functional Foods. 2021;**81**:1-13. DOI: 10.1016/j.jff.2021.104466

[4] Henry CJ. Functional foods. European Journal of Clinical Nutrition. 2010;**64**: 657-659. DOI: 10.1038/ejcn.2010.101

[5] Chen H, Ellett JK, Phillips R, Feng Y. Small-scale produce growers barriers and motivators to value-added business: Food safety and beyond. Food Control. 2021;**130**:1-13. DOI: 10.1016/j.foodcont.2021.108192

[6] Vattem DA, Maitin V. Functional Foods, Nutraceuticals and Natural Products. USA: DEStech Publications; 2016. p. 797

[7] FAO. The Future of Food and Agriculture. Alternative Pathways to 2050. Summary Version. Italy: FAO; 2018. p. 60

[8] Fernandes L, Saraiva JA, Pereira JA, Casal S, Rumalhosa E. Post-harvest technologies applied to edible flowers: A review. Food Reviews International. 2019;**35**:132-154. DOI: org/10.1080/87559129.2018.1473422

[9] Guiner PF, Florenca SG, Moya ku, Anjos O. Edible flowers, old tradition or new gastronomic trend: A first look at consumption in Portugal versus Costa Rica. Food. 2020;**9**:1-15. DOI: org/10.3390/foods9080977

[10] Rob O, Mlcek J, Jurikova T, Neugebauerova J, Vaskova J. Edible flowers – E new promising source of mineral elements in human nutrition. Molecules. 2012;**17**:6672-6683. DOI: 10.3390/molecules17066672

[11] Huis A. Edible insects are the future? The Proceedings of the Nutrition Society. 2016;**75**:1-60. DOI: 10.1017/S0029665116000069

[12] Delvendahl N, Rumpold BA, Langen N. Edible insects as food–insect welfare and ethical aspects from a consumer perspective. Insects. 2022;**13**: 1-13. DOI: 10.3390/insects13020121

[13] Kim TK, Yoong HI, Kim YB, Kim HW, CHoi YS. Edible insects as a protein source: A review of public perception, processing technology, and research trends. Food Science of Animal Resources. 2019;**39**:521-540. DOI: 10.5851/kosfa.2019.e53

[14] Ranga Rao A, Ravishankar GA. Algae as source of functional ingredients for health benefits. Agricultural Research and Technology. 2018;**14**:1.4. DOI: 10.19080/ARTOAJ.2018.14.5559

[15] Wells ML, Potin P, Craigie JS, Raven JA, Merchant SS, Helliwell KE, et al. Algae as nutritional and functional food sources: Revisiting our understanding. Journal of Applied Phycology. 2017;**29**:949-982. DOI: 10.1007/s10811-016-0974-5

[16] Prasad S, Rathore H, Sharma S, Yadav AS. Medicinal mushrooms as a source of novel functional food.

International Journal of Food Science, Nutrition and Dietetics. 2015;**4**:221-225. DOI: 10.19070/2326-3350-1500040

[17] Waktola G, Temesgen T. Application of mushroom as food and medicine. Advances in Biotechnology and Microbiology. 2018;**11**:97-101. DOI: 10.19080/AIBM.2018.11.555817

[18] Reinheimer MA, Mussati S, Scenna NJ. Influence of product composition and operating conditions on the unsteady behavior of hard candy cooling process. Journal of Food Engineering. 2010;**101**(4):409-416. DOI: 10.1016/j.jfoodeng.2010.07.029

[19] Hartel RW, von Elbe JH, Hofberger R. Hard candy. In: Confectionery Science and Technology. Cham: Springer; 2018. pp. 211-244. DOI: 10.1007/978-3-319-61742-8_8

[20] Marinelli V, Lucera A, Incoronato AL, Matteo LM, Nobile AD, Conte A. Strategies for fortified sustainable food: The case of watermelon-based candy. Journal of Food Science and Technology. 2021; **58**(3):894-901. DOI: 10.1007/s13197-020-04603-2

[21] Cornejo-Ramírez YI, Martínez-Cruz O, Del Toro-Sánchez CL, Wong-Corral FJ, Borboa-Flores J, Cinco-Moroyoqui FJ. The structural characteristics of starches and their functional properties. CyTA Journal of Food. 2018;**16**(1):1003-1017. DOI: 0.1080/19476337.2018.1518343

[22] Figiel A, Tajner-Czopek A. The effect of candy moisture content on texture. The Journal of Foodservice. 2006;**17**(4):189-195. DOI: 10.1111/j.1745-4506.2006. 00037.x

[23] Sahin S, Sumnu SG. Physical Properties of Foods. New York: Springer; 2006. p. 257

[24] Hlaváč P. Rheologic Properties of Food Materials [Dissertation Thesis]. Nitra: Slovak University of Agriculture in Nitra; 2010

[25] Figura LO, Teixeira AA. Food Physics, Physical Properties – Measurement and Applications. Berlin: Springer; 2007. p. 550

[26] Rao MA, Rizvi SSH, Datta AK. Engineering Properties of Foods. 3rd ed. LLC: Taylor & Francis Group; 2005. p. 738

[27] Severa L, Nedomová Š. Physical and Mechanical Properties of Foods (in Czech). Brno: Mendelova univerzita v Brně; 2011. p. 116

[28] Hlaváčová Z. Physical Properties of Foods (in Slovak). Nitra: SPU v Nitre; 2019. p. 146

[29] Steffe JF. Rheological Method in Food Process Engineering. 2nd ed. East Lansing: Freeman Press; 1996. p. 428

[30] Lu X, Hou L, Zhang L, Tong Y, Zhao G, Cheng ZY. Piezoelectric-excited membrane for liquids viscosity and mass density measurement. Sensors and Actuators A: Physical. 2017;**261**:196-201. DOI: 10.1016/j.sna.2017.05.014

[31] Xu L, Wu K, Han R, Sui Y, Huang C, Huang W, et al. Visual detection of viscosity through activatable molecular rotor with aggregation-induced emission. Spectrochimica Acta Part A. 2021;**261**:120016. DOI: 10.1016/j.saa.2021.120016

[32] Hashim IB, Khalil AH, Afifi HS. Quality characteristics and consumer acceptance of yogurt fortified with date fiber. Journal of Dairy Science. 2009; **92**(11):5403-5407. DOI: 10.3168/jds.2009-2234

[33] Gómez-Díaz D, Navaza JM, Quintáns-Riveiro LC. Rheological behaviour of Galician honeys. European Food Research and Technology. 2006; **222**:439-442. DOI: 10.1007/S00217-005-0120-0

[34] Kayacier A, Yüksel F, Karaman S. Dynamic mechanical spectra of selected Turkish honeys: Construction of predictive models for complex viscosity using two different nonlinear Modeling techniques. International Journal of Food Properties. 2014;**17**(1):93-110. DOI: 10.1080/10942912.2011.614985

[35] Cohen I, Weihs D. Rheology and microrheology of natural and reduced-calorie Israeli honeys as a model for high-viscosity Newtonian liquids. Journal of Food Engineering. 2010; **100**(2):366-371. DOI: 10.1016/j.jfoodeng.2010.04.023

[36] Abu-Jdayil B, Ghzawi AA-M, Al-Malah KIM, Zaitoun S. Heat effect on rheology of light- and dark-Colored honey. Journal of Food Engineering. 2002;**51**:33-38

[37] Afonso MJ, Magalhães M, Fernandes L, Castro M, Ramalhosa ECD. Temperature effect on rheological behavior of Portuguese honeys. Polish Journal of Food and Nutrition Sciences. 2018;**68**(3):217-222. DOI: 10.1515/pjfns-2017-0030

[38] Karasu S, Toker OS, Yilmaz MT, Karaman S, Dertli E. Thermal loop test to determine structural changes and thermal stability of creamed honey: Rheological characterization. Journal of Food Engineering. 2015;**150**:90-98. DOI: 10/1016/j.foodeng.2014.10.004

[39] Díaz-Ocampo R, Sánchez R, Franco JM. Rheology of commercial and model borojó jam formulations. International Journal of Food Properties. 2014;**14**(4):791-805. DOI: 10.1080/10942912.2012.665406

[40] Espinosa-Muñoz L, Renard CMGC, Symoneaux R, Biau N, Cuvelier G. Structural parameters that determine the rheological properties of apple puree. Journal of Food Engineering. 2013;**119**: 619-626. DOI: 10.1016/j.jfoodeng.2013.06.014

[41] Razavi SMA, Karazhiyan H. Rheological and textural characteristics of date paste. International Journal of Food Properties. 2012;**15**(2):281-291. DOI: 10.1080/10942912.2010.483615

[42] Qi B, Zhang Q, Sui X, Wang Z, Li Y, Jiang L. Differential scanning calorimetry study – Assessing the influence of composition of vegetable oils on oxidation. Food Chemistry. 2016; **194**:601-607. DOI: 10.1016/j.foodchem.2015.07.148

[43] Kumbár V, Nedomová Š, Ondrušíková S, Polcar A. Rheological behaviour of chocolate at different temperatures. Potravinarstvo Slovak Journal of Food Sciences. 2018;**12**(1): 123-128. DOI: 10.5219/876

[44] Liu F, Cao X, Wang H, Liao X. Changes of tomato powder qualities during storage. Powder Technology. 2010;**204**:159-166. DOI: 10.1016/J.POWTEC.2010.08.002

[45] Torbica A, Belović M, Mastilović J, Kevrešan Ž, Pestorić M, Škrobot D, et al. Nutritional, rheological, and sensory evaluation of tomato ketchup with increased content of natural fibres made from fresh tomato pomace. Food and Bioproducts Processing. 2016;**98**: 299-309

[46] Chandrapala J, Oliver C, Kentish S, Ashokkumar M. Ultrasonics in food processing – Food quality assurance and

food safety. Trends in Food Science and Technology. 2012;**26**:88-98. DOI: 10.1016/J.TIFS.2012.01.010

[47] Thomas MJ, Bramblett KA, Green BD, West KN. Thermophysical and absorption properties of brominated vegetable oil. Journal of Molecular Liquids. 2015;**211**:647-655. DOI: 10.1016/j.molliq.2015.07.060

[48] Alcântara LAP, Fontan RCI, Bonomo RCF, Souza EC Jr, Sampaio VS, Pereira RG. Density and dynamic viscosity of bovine milk affect by temperature and composition. International Journal of Food Engineering. 2012;**8**(1):556-568. DOI: 10.1515/1556-3758.1860

[49] Oguntunde AO, Akintoye OA. Measurement and comparison of density, specific heat and viscosity of cow's milk and soymilk. Journal of Food Engineering. 1991;**13**:221-230. DOI: 10.1016/0260-8774(91)90028-Q

[50] Sharoba AM, Senge B, El-Mansy HA, Bahlol HEIM, Blochwitz R. Chemical, sensory and rheological properties of some commercial German and Egyptian tomato ketchups. European Food Research and Technology. 2005;**220**: 142-151. DOI: 10.1007/s00217-004-0981-7

[51] Karaman S, Yilmaz MT, Cankurt H, Kayacier A, Sagdic O. Linear creep and recovery analysis of ketchup–processed cheese mixtures using mechanical simulation models as a function of temperature and concentration. Food Research International. 2012;**48**: 507-519. DOI: 10.1016/j.foodres.2012.05.016

[52] Pomeranz Y, Meloan CE. Thermal analysis of foods. In: Food Analysis. Boston: Springer MA; 1994. pp. 555-563. DOI: 10.1007/978-1-4615-6998-5_32

[53] Joardder MUH, Karim MA, Kumar C. Better understanding of food material on the basis of water distribution using thermogravimetric analysis. In: Proceeding of the International Conference on Mechanical, Industrial and Materials Engineering 2013 (ICMIME); 1–3 November 2013. Rajshahi, Bangladesh: RUET; 2013. DOI: 10.13140/RG.2.1.1508.8807

[54] Orsat V, Yang W, Changrue V, Raghavan GSV. Microwave-assisted drying of biomaterials. Food and Bioproducts Processing. 2007;**85**(3): 255-263. DOI: 10.1205/fbp07019

[55] Fraga LG, Silva J, Teixeira S, Soares D, Ferreira M, Teixeira J. Thermal conversion of pine wood and kinetic analysis under oxidative and non-oxidative environments at low heating rate. Proceedings. 2020;**58**:23. DOI: 10.3390/WEF-06921

[56] Wagner M. Thermal Analysis in Practice, Collected Applications. München: Carl Hanser Verlag, Mettler Toledo; 2009. p. 347

[57] Parniakov O, Bals O, Barba FJ, Mykhailyk V, Lebovka N, Vorobiev E. Application of differential scanning calorimetry to estimate quality and nutritional properties of food products. Critical Reviews in Food Science and Nutrition. 2018;**58**(3):362-385. DOI: 10.1080/10408398.2016.1180502

[58] Thermal Analysis of Foods [Internet]. Available from: http://people.umass.edu/~mcclemen/581Thermal.html. [Accessed: 2018-04-08]

[59] Halim AH, Selemat J, Mirhosseini S, Hussain N. Sensory preference and bloom stability of chocolate containing cocoa butter substitute from coconut oil. Journal of the Saudi Society of Agricultural Sciences. 2017;**18**(4):

443-448. DOI: 10.1016/j.
jssas.2018.02.005

[60] Costas G. Biliaderis, differential scanning calorimetry in food research - a review. Food Chemistry. 1983;**10**(4): 239-265. DOI: 10.1016/0308-8146(83) 90081-X

[61] Beckett ST. The Science of Chocolate. 2nd ed. Cambridge: Royal Society of Chemistry; 2008. p. 240. DOI: 10.1039/9781847558053-FP005

[62] Nelson SO, Trabelsi S. Use of material dielectric properties in agricultural applications. Journal of Microwave Power and Electromagnetic Energy. 2016;**50**:237-268. DOI: 10.1080/ 08327823.2016.1247235

[63] Hlaváčová Z. Low frequency electric properties utilization in agriculture and food treatment. Research in Agricultural Engineering. 2003;**49**(4):125-136. DOI: 10.17221/4963-RAE

[64] Jha SN, Narsaiah K, Basediya AL, Sharma R, Jaiswal P, Kumar R, et al. Measurement techniques and application of electrical properties for nondestructive quality evaluation of foods - a review. Journal of Food Science and Technology. 2011;**48**(4):387-411. DOI: 10.1007/s13197-011-0263-x

[65] Guo W, Tiwari G, Tang J, Wang S. Frequency, moisture and temperature-dependent dielectric properties of chickpea flour. Biosystems Engineering. 2008;**101**:217-224. DOI: 10.1016/j. biosystemseng.2008.07.002

[66] Nakonieczna A, Paszkowski B, Wilczek A, Szypłowska A, Skierucha W. Electrical impedance measurements for detecting artificial chemical additives in liquid food products. Food Control. 2016;**66**:116-129. DOI: 10.1016/j. foodcont.2016.01.044

[67] Sharma A, Bansod BK, Thakur R. Development of moisture prediction model for tea using electrical impedance spectroscopy. International Journal of Advanced Engineering Research and Science. 2017;**4**(2):38-43. DOI: 10.22161/ ijaers.4.2.8

[68] Cataldo A, Piuzzi E, Cannazza G, De Benedetto E, Tarricone L. Quality and anti-adulteration control of vegetable oils through microwave dielectric spectroscopy. Measurement. 2010;**43**: 1031-1039. DOI: 10.1016/j. measurement.2010.02.008

[69] Abbas Z, Yeow YK, Shaari AH, Khalid K, Hassan J, Saion E. Complex permittivity and moisture measurements of oil palm fruits using an open-ended coaxial sensor. IEEE Sensors Journal. 2005;**5**(6):1281-1287. DOI: 10.1109/ JSEN.2005.859249

[70] Nik Pauzi NNP, Ahmad Hazmi AS, Abdul Aziz H, Abu Bakar Z, Idris Z. Frequency response and activation energy of palm-fatty acids at microwave region. International Journal of Hydrogen Energy. 2017;**42**(32): 20444-20452. DOI: 1016/j. ijhydene.2017.06.185

[71] Bertemes-Filho P, Valicheski R, Pereira R, Paterno A. Bioelectrical impedance analysis for bovine milk: Preliminary results. Journal of Physics. 2010;**224**(1):012133. DOI: 10.1088/ 1742-6596/224/1/012133

[72] Blahovec J. Electrical and optical properties of agricultural products and foods (in Czech). Československý časopis pro fyziku. 2017;**67**(2):60-90

[73] Martin O, Qin BL, Chang FJ, Barbosa-Cánovas GV, Swanson BG. Inactivation of Escherichia coli in skim milk by high intensity pulsed electric fields. Journal of Food Process

Engineering. 1997;**20**:317-336. DOI: 10.1111/j.1745-4530.1997.tb00425.x

[74] Guo W, Lin B, Liu D, Zhu X. A novel technique on determining water content in milk using radio-frequency/ microwave dielectric spectroscopy and chemometrics. Food Analytical Methods. 2017;**10**:3781-3789. DOI: 10.1007/ s12161-017-0946-7

[75] Pattee HE. Evaluation of Quality of Fruits and Vegetables. New York: Springer Science + Business Media; 2014. p. 428

[76] Caicedo-Eraso CJ, Diaz-Arango OF, Osorio-Alturo A. Electrical impedance spectroscopy applied to quality control in the food industry. Ciencia & Tecnología Agropecuaria. 2020;**21**(1): 100-119. DOI: 10.21930/rcta.vol21_ num1_art:951

[77] Guo W, Nelson SO, Trabelsi S, Kays S. 0–1800 MHz dielectric properties of fresh apples during storage. Journal of Food Engineering. 2007;**83**: 562-569. DOI: 10.1016/j. jfoodeng.2007.04.009

[78] Watanabe T, Ando Y, Orikasa T, Kasai S, Shiina T. Electrical impedance estimation for apple fruit tissues during storage using Cole–Cole plots. Journal of Food Engineering. 2018;**221**:29-34. DOI: 10.1016/j.jfoodeng.2017.09.028

[79] Łuczycka D, Czubaszek A, Fujarczuk M, Pruski K. Dielectric properties of wheat flour mixed with oatmeal. International Agrophysics. 2013;**27**:175-180. DOI: 10.2478/ v10247-012-0083-x

[80] Skierucha W, Wilczek A, Szypłowska A. Dielectric spectroscopy in agrophysics international agrophysics. International Agrophysics. 2012;**26**: 187-197. DOI: 10.2478/v10247-012-0027-5

[81] El Khaled D, Castellano NN, Gázque JA, Perea-Moreno AJ, Manzano-Agugliaro F. Dielectric spectroscopy in biomaterials: Agrophysics. Materials. 2016;**9**(310):1-26. DOI: 10.3390/ ma9050310

Chapter 5

Context-Specific Food-Based Strategies for Improving Nutrition in Developing Countries

Jofrey Raymond

Abstract

Viable food approaches for achieving nutrient needs in underdeveloped countries are not well documented. The existing evidence indicates that one out of three people is facing single or multiple forms of malnutrition globally, in which the highly affected sections of the population are children and women from less developed countries. Economic losses, which result from undernutrition are between 3% and 16% of the GDP in the majority of poor countries. This problem is far bigger than what the government and donors can tackle alone. Thus, a new strategy, which is donor-independent, is required to address the problem of undernutrition in developing countries. In this chapter, we report on a food approach that is context-specific for grappling with malnutrition problems in low-income countries. The approach employs the model which encompasses public and private sectors to allow cost-sharing and productivity gains in tackling malnutrition in under-resourced countries. The model urges all stakeholders to consider consumers' views, which are often overlooked, and properly engross them as key players.

Keywords: undernutrition, food-based approach, public-private hybrid-delivery model, context-specific approach, linear programming, low-income countries

1. Introduction

Suitable food approaches for attaining nutrient requirements in resource-poor communities are not well documented [1, 2]. Existing evidence indicates that not less than two billion individuals worldwide lack vital nutrients needed by the human body for its growth and proper functioning. Also, more than 0.84 billion people have no sufficient food to eat, whereas 1.4 billion people are morbidly obese or overweight. Overall, half of the world's population suffers from malnutrition [3]. The world's population is estimated to reach nine billion people by 2050, and these will require nutritious diets for healthy lives [4]. Therefore, it is essential to ensure that sustainable food approaches are initiated to achieve the global demands for nutrient-rich foods since the state of malnutrition is anticipated to be worse, if not addressed [1].

Although Sub-Saharan Africa has widely been regarded as the basket of foods rich in essential nutrients, malnourishment is still a great challenge, particularly, among children and women of reproductive age. More than 0.056 billion children, for example, are reported to be stunted due to chronic malnutrition, 0.03 billion are underweight, greater than 0.015 billion are wasted, nearly 50% of children are susceptible to blindness as a result of vitamin A deficiency, and four million low-birth-weight babies are born annually in the region [3]. In that regard, maternal undernutrition in Africa is considerably larger than in other parts of the world [1]. Nearly 5–20% of African women, for instance, have a low body mass index (BMI) attached to chronic hunger. Prevalence of anaemia varies from 21 to 80% of females across the African region. Moreover, vitamin A, zinc, and iron deficiencies are more prevalent in the continent [5]. To a great degree, maternal malnutrition accounts for millions of infancy morbidity, and death every year in less developed countries [6].

In Tanzania, for instance, more than 35.5% of all children below 2 years are reported to be chronically undernourished [7]. Also, nutritional deficiencies are widespread. For example, the prevalence of vitamin A deficiency, iron deficiency, and anaemia among children aged below 2 years are reported to be 33%, 42%, and 73%, respectively [8]. Equally, according to the Tanzania Demographic and Health Survey (TDHS) report of 2010, more than 30% of women of reproductive age had iron deficiency, 36% had iodine deficiency, 37% had vitamin A deficiency, and 40% had anaemia [8]. As a consequence, undernutrition affects the nutritional status of potential mothers, thereby leading to long-term negative health and economic consequences for their offspring [1].

2. Consequences of undernutrition

The adverse effects of undernutrition are pervasive, long-lasting, and irreversible, particularly when it happens during the critical first 1000 days of an individual's life commencing at the onset of pregnancy and stretching through his/her second year after birth [1]. Undernutrition intensifies not only the risks of morbidity and death but also weakens the initial physical and intellectual development of survivors [6]. The short and long-term adverse effects of malnutrition are expounded below:

2.1 Increased risks of mortality

More than 45% of avertable child mortalities are attributed to undernutrition [6]. A report by Global Panel on agriculture and food systems for nutrition, for instance, showed that severely malnourished children are nine times more likely to die than well-nourished children [9]. Nevertheless, the underlying undernourishment is usually disguised by an immediate cause of mortality that frequently exhibits itself as an infection [1]. Maternal death has a substantial causal association with undernutrition, of which the most common maternal death linked to the perinatal period manifests itself either through pregnancy snags or death at childbirth [1]. Generally, undernutrition insidiously minimises the natural life span of women [10, 11]. Thus, evertable death signifies a loss of human resources which distress households by subjecting children to more helplessness [12, 13], and the community at large by debasing social interactions which were built by mothers [1].

2.2 Intensified reoccurrance of illnesses

Over many years, it has been acknowledged that malnutrition causes nutrient deficiency-related diseases. For example, protein deficiency causes kwashiorkor and overall poor physical and mental development; vitamin A deficiency results in impaired vision; mineral deficiency, for example, iron causes iron-deficiency anaemia [14]. The devious effects of malnourishment are a general increase in vulnerability to infection, the severity of resultant ailments and dwindled outcomes of recovery periods and the high cost of handling the illness. In less developed countries, families bear a significant portion of the treatment costs, and to a certain extent, the national health care or health insurance companies. This causes a drain on the family budget, nation's resources, and the diversion of limited resources to other endeavours [1]. As described in the Global Panel Report, for instance, a full course of remedies for saving the life of a sternly wasted child costs between US$ 100 and US$ 200 per child [9], which is more or less two times the average family earnings per month in affected communities. It is possible to save about USD18 on every dollar invested in thwarting chronic undernourishment. In the USA, where there is a crisis of obesity, the per capita healthcare expenses are 80% higher for perversely overweight people than those with healthy normal weight [9]. While the healthcare system in the USA noticeably differs from those in underdeveloped nations, it is obvious that as some sections of the population turn overweight (particularly among the escalating urban poor), the fiscal burden will be exacerbated by a triple helix of undernutrition, infection, and obesity [1].

2.3 Delayed physical development

Sub-optimal physical development, often accompanied by life-long exposure to diseases, weakens economic output through reduced labour efficiency or absence from work. While the losses to individuals from malnutrition in poor countries have been projected at 10% or more of lifetime incomes [9, 15], the cost of productivity foregone in resources-poor countries because of undernourishment has been estimated at 3–16% of country's GDP [9, 16]. This lost income is equivalent to or surpasses the amount of GDP spent on agriculture or education, and this is likely to be more than what African states have devoted themselves to spending on scientific research and development [1]. It is useful to point out that the most developed countries like the United States, employment malingering associated with obesity causes lost output equal to US$ 4.3 billion per annum, and costs firms US$ 506 yearly per corpulent employee [9].

2.4 Poor cognitive development

Undernutrition from conception to 2 years of age and extending through school age and adolescence leads to a series of damages beginning with reduced cognitive development, delayed school starting, and poor educational achievements. As a result, the affected individuals miss opportunities related to lost employment and social relations during the course of life [1, 6, 9]. A longitudinal study conducted in Guatemala which involved 2392 children below 8 years, for example, revealed that malnourished children aged 6 years had a relatively high risk of losing the equivalent of four grades of schooling because of poor cognitive abilities [17]. Furthermore, a comparable long-term study piloted in Western Tanzania indicated that undernourished children are more likely to delay entry into school and perform poorer than well-nourished children of similar age [18].

Overall, malnutrition carries a substantial economic burden for the affected individuals, families, and communities as well as national and global economies [1].

The global analysis on nutrition economics indicates that a combined burden of malnutrition, micronutrient deficiencies, and obesity cost the world economy up to US$ 3.5 trillion [3, 9]. This level of economic burden is the main obstacle to efforts that are done by many governments to alleviate poverty and attain key development goals stipulated in the Sustainable Development Goals (SDGs) [9, 19]. These consequences emphasise the necessity of establishing suitable nutritional interventions for mitigating the problem of undernutrition in resource-poor communities [1].

3. Potential causes of undernutrition in resource-poor countries

It is clearly understood that undernutrition among individuals in poverty-stricken families and societies is caused by insufficient intake of nutritious foods. Nonetheless, this hypothesis is a generalisation because the insufficiency of food intake is itself caused by underlying factors, including poverty as chief among them [1]. Another aspect of the underlying causes of undernourishment is linked to eating habits that are monotonous, mainly cereal-based intakes, which tend to have low dietary diversity, low nutrient density, and meagre micronutrient bioavailability [20]. Also, for a long time, it has been widely acknowledged that most poor families concentrate on tackling the challenge of hunger and normally forget to take into account the quality and quantity of essential nutrients in foods [21]. A study on the formulation of evidence-based dietary references for women and children, for instance, revealed that foods rich in carbohydrates are more frequently consumed than locally available micronutrient-rich foods in Guatemala [22]. Likewise, an analysis of data from the TDHS indicated that highly sugared foods are more frequently consumed than foods with high content of essential micronutrients, suggesting that eating habits among toddlers in Tanzania are certainly poor [23]. Most Tanzanian families trade nutrition for food, meaning that they would rather sell their high-value nutrient-dense foods in exchange for bulky low-nutrient dense cereal, sugar, and other foodstuffs [1]. Undoubtedly, this is mainly because most of people in Tanzania have no enough nutritional education about the health benefits of nutrient-rich foods. This evidence indicates that it is necessary to have a robust nutrition intervention that would help households in disadvantaged communities to make informed choices on what, when, and how to eat.

4. Existing approaches for addressing malnutrition in developing countries

For the past 50 years, initiatives to addressing malnourishment in underprivileged communities have focused mainly on three areas of intervention. The interventions include (1) nutritional education to the public, (2) provision of micronutrients via salt iodisation, vitamin A supplements etc., and (3) the treatment of severe acute undernutrition through clinical intensive care fortification [1, 2].

Public nutrition education intervention aims to improve the consumption of available nutrient-dense foods for the nutritional benefits of disadvantaged groups. This intervention intends to change the purchasing, preparations, and eating behaviours to counter some deleterious food-related practices that tend to exist in diverse communities [1]. Precisely, nutrition education aims at creating changes in feeding habits based on the supposition that such changes will improve the nutritional status of the target community. For that matter, knowledge and attitude changes are expected to contribute to behavioural changes among individuals. In case the changes in behaviour do not positively impact the nutritional status of the target population, it might imply that the messages promoted in nutritional education were incorrect or that other causes were the major constraints to impaired growth [1, 24].

Fortification is normally done by adding micronutrient powder, ready-to-use therapeutic foods (RUTFs) and lipid-based nutrient supplements (LNS) in diets consumed in household. On the other hand, supplementation is usually done by providing a single nutrient (say vitamin A or D, iron) to infants and toddlers in the form of drops or syrups [1]. Regrettably, LNS and RUTFs cannot be afforded by many families in low-income countries and they are intended for use under clinical prescriptions only [25, 26]. Fortification and supplementation interventions have been based on the aids from donors who are always more reactive than proactive as their strategy has for a long time been on the treatment of severe acute malnutrition (SAM) rather than prevention of malnutrition [5, 27, 28]. Although these approaches are indispensable for mitigating life or death circumstances on an emergency basis, these efforts are generally considered to have been inadequate, leaving more than one-third of the global population to suffer from malnutrition [27].

5. Innovative approaches for meeting nutrient need in poor countries

At present, the WHO recommends the food-based strategy as a favoured long-term solution to addressing the problem of malnutrition in developing countries. The approach intends to enhance the nutritional quality through improved intake of local foods rich in essential nutrients [29]. Yet, for the intervention to work better, it must be specific to the culture and local context. The approach needs to be executed based on the acceptance, local availability, and affordability of the nutrient-rich foods in the community [30]. Studies from countries such as Cambodia, Guatemala, Indonesia, and Myanmar have shown that different vicinities may have diverse local foods with variable nutrient densities [7, 31, 32]. The fact that local foods can vary from one location to another highlights the need for examining how local diets in a specific locality can attain the nutritional goals of disadvantaged groups towards food-based interventions [1]. The local analysis helps nutrition stakeholders and practitioners resolve the obstacles that may be caused by factors like access and high prices of nutrient-rich foods, particularly in under-resourced settings [1].

As such, diet optimisation using linear programming (LP) is instrumental in formulating nutritious diets and providing sensible use of scarce high value nutrient-rich foods in a target population. LP is a mathematical tool used to optimise a linear objective function subject to a set of constraints [33]. Several objective functions can be selected, for instance, through minimising or maximising particular nutrients and the selected local food ingredients. The LP approach considers numerous factors including food costs and cultural factors in ensuring the smooth development of a reasonably priced and culturally acceptable diet [34]. The evidence from Kenya and Mozambique reveals how low-cost diverse diets can be attained from local nutrient-rich foods using linear programming. In Kenya, for example, the LP approach was used to model available wild foods to achieve nutritional goals for mothers and children aged below 2 years at a minimal cost [35]. In Mozambique, the LP technique was used to model local nutrient-dense foods to meet micronutrients necessities for children aged between 1 and 3 years [36]. Their LP diets formulated from local foods were reasonably affordable when compared to the cost of commonly used lipid-based nutrient supplements (LNS) [26]. These case studies confirmed the usefulness of food-based approaches that are context-specific and their ability to successfully tackle the problem of malnourishment in developing countries [1].

6. Necessities for the success of food-based approaches

For a food-based intervention to succeed, programme executers need to focus on developing foodstuffs that can meet consumer preferences and experiences, especially where obstacles are too deeply rooted to be altered in the short-term by educational tactics [1]. Foods need to be nontoxic and rich in the most deficient micronutrients, and they should be consumed as per intake recommendations [37]. In some situations, one must ensure that the product is acquired by dispensing it directly to the target population. Nevertheless, in this case, investigators and programme executors cannot guarantee the actual consumption of the product by the intended customers since the target beneficiaries may not necessarily consume the presented food product as planned [1]. This may partly be attributed by the circumstance that consumers dislike the product or are not aware that the product is intended for particular persons, or consumers feel anxious to share the food with other members inside or outside their households [37]. In view of that, prior to introducing a food-based intervention, it is imperative to carry out the analysis of total nutrient intakes, including that of a specific product and the whole diet. Also, to ensure consumption, ingredients of interest need to achieve consumer preferences, preparations, and consumption experiences. Equally, one requires to know the purchasing power of consumers and their willingness to pay for the product [37, 38]. In the same vein, in order to arouse consumption as planned and limit the sharing of foods in communities, clear and well-tailored communications should be developed and delivered to the target population [37, 39].

After identifying the characteristics of target consumers such as consumers' purchasing power and behaviours, eating manners, and food consumption patterns, products or formulations can be developed and tested at a small scale to establish consumer preferences and viable affordability. Data from such a study may act as a basis for moving into a larger micro-testing [3]. Once consumer preferences and purchasing power have been resolved, one needs to develop a well-tailored delivery system to ensure the sustainable distribution of nutritious food products to the target population [1].

7. Proposed delivery system for nutritious food products in developing countries

A sustainable delivery model for effective distribution of nutritious products to the target population is needed in resource-poor settings. Studies have shown that in real-life situations, target consumers can be reached through health care, market networks, and community-based channels. However, according to de Pee [37], the choice of a channel or combination of channels depends on the context, type of product, consumers' purchasing power and their preferences, likelihood for behavioural change, available platforms and assurance among the target users [37].

Segre and the team, for instance, carried out research that assessed the 'actual' consumers' willingness to pay for nutrient-dense foodstuffs among urban people in Ethiopia. The study's objectives were to: (1) explore the size of the market and purchasing power of individuals, (2) determine the local manufacturer's business case, and (3) establish the required attributes of the distribution channel. The study found that although the majority of consumers preferred to buy the food product and that 25% of participants were ready to pay an unsubsidised price, yet, most of them were not used to buying packaged foods, and hence, the total market size was small [40]. The study's findings suggest that to reach 75% of urban consumers who were not willing to pay the unsubsidised price for the products in Ethiopia would need a departure from the business as usual approach. It would possibly need a delivery model which embraces a public-private mix, which allows cost-sharing and efficiency gains [28].

The hybrid public-private delivery model is being proposed as a more profound and beneficial channel of distribution for nutritious foods over other channels in developing countries. This delivery model crafts an environment that forces key nutrition players and other practitioners to seriously take into account consumers' views that are often ignored, and truly engross them as key stakeholders [28, 41]. In this model, the public sector is responsible for providing conducive environments for investments, infrastructure, and legal frameworks needed to support the private sector (both inside and outside the country) to engage more extensively in executing the nutrition business [1]. The public sector can also provide a direct subsidisation, which plays a key role in lowering the retail price of nutritious products. Likewise, the public sector can assist in reducing production costs by eliminating unnecessary trade obstacles and tariffs for key production inputs. Buyers from the public sector also have an important role to play in minimising the uncertainty related to the demand for potential nutritional products. Public sector procurers and purchasers can also sign credible and long-term contracts to stimulate private sector investments in resource-poor countries [41]. Eventually, this would ensure that there is a sustainable supply of nutritional products and formulations to the target population [1].

Given that the problem of undernutrition is far bigger than what the donor-driven and government intervention programmes can tackle alone, a context-specific food-based approach that would allow cost-sharing and efficiency gains in low-income countries is needed [1]. This approach is grounded on the fact that a mixed hybrid public-private delivery model is not donor-dependent and hence, can assure and enhance a sustainable delivery of nutritious nutrient-dense foodstuffs to the target consumers. In that way, the hybrid public-private delivery model guarantees a sustainable supply of nutritious commodities to the target population, including those who cannot manage to pay for the unsubsidised price. This delivery model considers retail channels as a potential delivery platform for sharing the cost burden with consumers. The model provides an avenue for key nutritional actors to leverage a private demand by marketing nutritious foodstuffs in the famous retail channels or outlets at a subsidised price or by charging fees at the distribution health care centres [1]. Nevertheless, market demand will set a basis not only for designing, testing, and scaling up production and delivery options but also for tackling cost-sharing issues in less developed countries.

8. General conclusions

Food-based nutrition intervention options that are context-specific can be used to address malnutrition problems in developing countries. Based on the available analysis and our recent pilot study on the nutritional framework in Tanzania [1], we recommend three key food-based nutrition intervention options in resource-poor settings.

The first intervention option is by providing new avenues for increasing the production and wide consumption of rarely consumed micronutrient-dense foods in the target population. This will ensure the availability and accessibility of nutrient-rich foods in the target population. The approach may work better in areas with high ecosystem productivity. Apart from increased production of nutrient-dense foods, this model can provide economic incentives for caregivers to generate household income that can be spent on health care and the purchase of diverse micronutrient-rich foods at the household level.

The second intervention option is to establish the hybrid public-private delivery model that enables cost-sharing and efficiency gains among stakeholders. This model is more suitable in areas with relatively lower productivity ecosystems. In this model, the primary caregivers or consumers are truly engaged as key stakeholders. The role of the public sector in this hybrid model is to provide investments, infrastructure and legal frameworks for the private sectors to engage themselves in the production and distribution of nutrient-dense foods or formulations. The public sector in this hybrid model can also lower production costs through the reduction of trade barriers and tariffs for basic production inputs. Direct subsidies from the public sector can also lower the price of the formulation and hence increase the pool of purchasers from low-income households. However, one needs to persuade the governments in developing countries to understand the economic gain of investing in nutrition through direct subsidies. This can be compared with the costs governments are incurring in treating nutrition-related diseases.

The third intervention option is to empower mothers or primary caregivers by training them on skills for formulating nutritious products and diets at the household level using locally available foods and resources. The economic returns for this model are likely to be relatively high and stable because it minimises costs associated with labour, overhead costs, production, packaging, marketing, and distribution of the nutritional formulations. This approach can expand the pool of primary caregivers from low-income households to access nutritional solutions at a relatively lower cost.

Author details

Jofrey Raymond
The Nelson Mandela African Institution of Science and Technology (NM-AIST), Arusha, Tanzania

*Address all correspondence to: jofrey.raymond@gmail.com

References

[1] Raymond J, Kassim N, Rose JW, Agaba M. Context-specific food-based approach for ensuring nutrition security in developing countries: A review. International Journal of Food Sciences and Nutrition. 2017;**0**:1-7

[2] Reinhardt K, Fanzo J. Addressing chronic malnutrition through multi-sectoral, sustainable approaches: A review of the causes and consequences. Frontiers in Nutrition. 2014;**1**(August):1-11

[3] FAO/WHO. New Business Models to Help Eliminate Food and Nutrition Insecurity: Roadmap for Exploration. Rome, Italy: Food and Agriculture Organization of the United Nations; 2013

[4] Food and Agriculture Organization of the United Nations. How to Feed the World in 2050: High Level Expert Forum. Rome, Italy: Food and Agriculture Organization of the United Nations; 2009

[5] Lartey A. Maternal and child nutrition in Sub-Saharan Africa: Challenges and interventions. Proceedings of the Nutrition Society. 2008;**67**(01):105-108

[6] Black RE, Victora CG, Walker SP, Bhutta ZA, Christian P, De Onis M, et al. Maternal and child undernutrition and overweight in low-income and middle-income countries. The Lancet. 2013;*382*(9890):427-451

[7] Vitta BS, Benjamin M, Pries AM, Champeny M, Zehner E, Huffman SL. Infant and young child feeding practices among children under 2 years of age and maternal exposure to infant and young child feeding messages and promotions in Dar es Salaam, Tanzania. Maternal & Child Nutrition. 2016;**12**:77-90

[8] TDHS. Tanzania Demographic and Health Survey 2010. Maryland, USA: National Bureau of Statistics Dar Es Salaam, Tanzania ICF Macro Calverton; 2010. pp. 1-482

[9] Global Panel. The Cost of Malnutrition: Why Policy Action is Urgent. London, UK: Global Panel on Agriculture and Food Systems for Nutrition; 2016

[10] Black RE, Allen LH, Bhutta ZA, Caulfield LE, de Onis M, Ezzati M, et al. Maternal and child undernutrition: Global and regional exposures and health consequences. The Lancet. 2008;**371**(9608):243-260

[11] Victora CG, Adair L, Fall C, Hallal PC, Martorell R, Richter L, et al. Maternal and child undernutrition 2. Maternal and child undernutrition: Consequences for adult health and human capital. The Lancet. 2008;**371**:340-357

[12] Molla M, Mitiku I, Worku A, Yamin AE. Impacts of maternal mortality on living children and families : A qualitative study from Butajira, Ethiopia. Reproductive Health. 2015;**12**(Suppl. 1):1-9

[13] Moucheraud C, Worku A, Molla M, Finlay JE, Leaning J, Yamin AE. Consequences of maternal mortality on infant and child survival: A 25-year longitudinal analysis in Butajira Ethiopia (1987-2011). Reproductive Health. 2015;**12**(Suppl. 1):S4

[14] Müller O, Krawinkel M. Malnutrition and health in developing countries. 2005;**173**(3):2000-2003

[15] World Bank. Repositioning nutrition as central to development: A strategy for large-scale action. Directions in Development Series. Washington, D.C. 2006;**13**(9):272. DOI: 10.1596-978-0-8213-6399-7

[16] Hoddinott J. The economics of reducing malnutrition in Sub-Saharan Africa. Global Panel on Agriculture and Food Systems for Nutrition Working Paper. 2016

[17] Maluccio JA, Hoddinott J, Behrman JR, Martorell R, Quisumbing AR, Stein AD. The impact of improving nutrition during early childhood on education among Guatemalan adults. Economic Journal. 2009;**119**(537):734-763

[18] Alderman H, Hoogeveen H, Rossi M. Preschool nutrition and subsequent schooling attainment: Longitudinal evidence from Tanzania. Economic Development and Cultural Change. 2009;**57**:239-260

[19] SDG. The United Nations 2030 Agenda: 17 Sustainable Development Goals to Transform Our World. 2015

[20] Osendarp SJM, Broersen B, van Liere MJ, De-Regil L, Bahirathan L, Klassen E, et al. Complementary feeding diets made of local foods can be optimized, but additional interventions will be needed to meet iron and zinc requirements in 6- to 23-month-old children in low- and middle-income countries. Food and Nutrition Bulletin. 2016:1-27

[21] FAO. Human Nutrition in the Developing World. Rome, Italy: Food and Agriculture Organization of the United Nations; 1997

[22] FANTA. Development of Evidence-Based Dietary Recommendations for Children, Pregnant Women, and Lactating Women Living in the Western Highlands of Guatemala. Washington, DC: FHI 360/FANTA; 2014

[23] Huffman SL, Piwoz EG, Vosti SA, Dewey KG. Babies, soft drinks and snacks: A concern in low- and middle-income countries? Maternal and Child Nutrition. 2014;**10**(4):562-574

[24] Austin JE, Zeitlin MF. Nutrition Intervention in Developing Countries: An Overview. Cambridge, MA: Oelgeschlager, Gunn & Hain, Publishers, Inc; 1981

[25] Kodish SR, Aburto NJ, Nseluke Hambayi M, Dibari F, Gittelsohn J. Patterns and determinants of small-quantity LNS utilization in rural Malawi and Mozambique: considerations for interventions with specialized nutritious foods. Maternal & Child Nutrition. Jan 2017;**13**(1). DOI: 10.1111/mcn.12234

[26] Robertson A, Parlesak A, Greiner T, Kent G, Cattaneo A. Biodiverse food solutions to enhance complementary feeding. Maternal & Child Nutrition. 2016;**12**(3):639-640

[27] GAIN. Shaping markets to combat undernutrition. Rethinking approaches to scaling up the role of the private sector: Delivery models and early lessons (September). 2011

[28] Lybbert TJ. Hybrid public-private delivery of preventative lipid-based nutrient supplement products: Key challenges, opportunities and players in an emerging product space. SCN News. 2011;**39**:32-39

[29] WHO/UNICEF. Global strategy for infant and young child feeding. Report. 2003. pp. 1-30

[30] Vossenaar M, Knight FA, Tumilowicz A, Hotz C, Chege P, Ferguson EL. Context-specific complementary feeding recommendations developed using Optifood could improve the diets of breast-fed infants and young children from diverse livelihood groups in northern Kenya. 2016;**20**(6):971-983

[31] Fahmida U, Kolopaking R, Santika O, Sriani S, Umar J, Htet MK, et al. Effectiveness in improving knowledge, practices, and intakes of key problem nutrients of a complementary feeding intervention developed by using linear programming: Experience in Lombok, Indonesia. American Journal of Clinical Nutrition. 2015;**101**(3):455-461

[32] Hlaing LM, Fahmida U, Htet MK, Utomo B, Firmansyah A, Ferguson EL. Local food-based complementary feeding recommendations developed by the linear programming approach to improve the intake of problem nutrients among 12-23-month-old Myanmar children. 2016;**116**:16-26

[33] Luenberger DG, Yinyu Y. In: Hillier FS, editor. Linear and Non-linear Programming. 3rd ed. New York: Springer Science + Business Media, LLC; 2008

[34] Ferguson EL, Darmon N, Briend A, Premachandra IM. Food-based dietary guidelines can be developed and tested using linear programming analysis. Journal of Nutrition. 2004;**134**:951-957

[35] Termote C, Raneri J, Deptford A, Cogill B. Assessing the potential of wild foods to reduce the cost of a nutritionally adequate diet: An example from eastern Baringo District, Kenya. Food and Nutrition Bulletin. 2014;**35**(4):458-479

[36] Parlesak A, Geelhoed D, Robertson A. Toward the prevention of childhood undernutrition: Diet diversity strategies using locally produced food can overcome gaps in nutrient supply. Food and Nutrition Bulletin. 2014;**35**(2):191-199

[37] de Pee S. Special nutritious solutions to enhance complementary feeding. Maternal and Child Nutrition. 2015;**11**:i-viii

[38] Gradl C, Jenkins B. Tackling Barriers to Scale: From Inclusive Business Models to Inclusive Business Ecosystems. Harvard Kennedy School: Corporate Social Responsibility; 2011. pp. 1-36

[39] Kodish S, Rah JH, Kraemer K, de Pee S, Gittelsohn J. Understanding low usage of micronutrient powder in the Kakuma refugee camp, Kenya: Findings from a qualitative study. Food and Nutrition Bulletin. 2011;**32**(3):292-303

[40] Segrè J, Winnard K, Abrha TH, Abebe Y, Shilane D, Lapping K. Willingness to pay for lipid-based nutrient supplements for young children in four urban sites of Ethiopia. Maternal and Child Nutrition. 2015;**11**:16-30

[41] Lybbert TJ, Vosti SA, Adams KP, Guissou R. Household Demand for Child Micronutrient Supplementation in Burkina Faso. Berkeley: Center for Effective Global Action/University of California; 2016

Chapter 6

Pulsed Electric Fields as a Green Pretreatment to Enhance Mass Transfer from Grapes of Bioactive Molecules: Aromatic, Phenolic, and Nitrogen Compounds

Teresa Garde-Cerdán, Eva P. Pérez-Álvarez, Pilar Rubio-Bretón and Noelia López-Giral

Abstract

The objective of this chapter was to summarize the effects of four pulsed electric field (PEF) treatments on the chemical composition of three grape varieties. To this end, Graciano, Tempranillo, and Grenache grapes were destemmed and crushed and then were treated using a PEF continuous system. Phenolic and nitrogen compounds were analyzed by HPLC and volatile compounds by GC-MS. The results showed that the influence of PEF treatments on these bioactive molecules was different depending on the grape variety and PEF treatment applied. This non-thermal technology improved Grenache aromatic composition, but Tempranillo and Graciano volatile composition was not affected by PEF. The treatment with the highest time and energy was the most effective on the total stilbene extraction, greatly increasing the content of these compounds in all grape varieties. Moreover, all PEF treatments enhanced yeast assimilable nitrogen (YAN) and total amino acids of Grenache grape variety, while for Graciano and Tempranillo, the PEF treatments hardly affected its nitrogen compounds content. In conclusion, from the point of view of the chemical composition and taking into account the treatments used in this study, it can be concluded that PEF is an appropriate technology to improve the quality of Grenache variety.

Keywords: volatile compounds, stilbenes, free amino acids, must, PEF, varietal aroma, flavor, resveratrol, piceid, YAN

1. Introduction

Modern winemaking requires elaborating quality wines, but also beneficial to health. The winemaking industry, to be competitive, has to develop and to take new processes that allow to get these requirements maintaining the sensory quality at the highest possible level. Therefore, it is particularly important to have new technologies that allow to improve the processes and to optimize the quality. Pulsed electric field technology (PEF) is considered one of the non-thermal methods for inactivating microorganisms in foods and also enhances mass transfer by electroporation of the cytoplasmatic membranes [1, 2]. In this way, in the extraction of grape bioactive components, some studies have been carried out on the effect of this technology on the extraction of several compounds, most of which are focused on the study of phenolic compounds [3, 4]. In addition, studies using continuous semi-industrial systems are scarce [3–5]. The three families of compounds that most influence grape quality are: volatile compounds, phenolic compounds, and nitrogenous compounds. Grape volatile composition is one of the key parameters determining must and wine quality [6].

These volatile compounds are located in the grape both in the pulp and in the skin and depend fundamentally on the variety, cultural practices, soil type, and geographical place [7, 8]. They are responsible for the varietal or primary wine aroma and are composed of several hundreds of compounds of different chemical groups integrated as monoterpenoids, C13 norisoprenoids, and benzenoid compounds from the aroma of the grape [9–11]. Two groups of most odoriferous which give floral aroma are monoterpenoids and C13 norisoprenoids [12].

On the other hand, C6 compounds are the major group of volatile compounds formed in the pre-fermentative stage, and they can have a negative effect on wine quality due to their herbaceous odors [13]. Resveratrol is a stilbene that has been the most widely studied phenolic compound due to its beneficial properties attributed to it, such as cardioprotective capacity, antioxidant, anticancer, antidiabetic, neuroprotective, and anti-aging activities [14, 15].

In nature, resveratrol can be found in two isomeric forms, cis and trans. Also, the glycosylated form, known as piceid, is the most abundant [16, 17].

Nitrogen compounds are quantitatively the second most abundant compounds in grapes, after sugars. This fraction is present in different forms, ammonium, amino acids, peptides, and proteins [18]. The quantity and quality of these compounds, mainly ammonium and amino acids, determine the growth of yeast and the fermentation rate [19–23]. Moreover, esters, higher alcohols, volatile fatty acids, and carbonyls are important contributors to the wine fermentation bouquet [24]. These compounds principally arise as metabolites of yeast sugar and amino acids [25], and their formation is affected by the nitrogen compounds present in the initial must [26, 27]. Therefore, the study of the amino acids content of the grape juice is relevant to estimating the aromatic profile of wine [28].

For these reasons, this work aimed to study the composition of must and wine from grapes treated by different PEF conditions using a continuous system of pilot scale.

2. Materials and methods

2.1 Graciano, Tempranillo, and Grenache grape samples

The study was carried out with three red grape varieties from the D.O.Ca. Rioja: Graciano, Tempranillo, and Grenache. A total of 400 kg of grapes of each variety were harvested at their optimum technology maturity. Then grapes were processed as in industry, were destemmed, crushed, and sulphited with 70 mg/kg SO_2. A total 10 stainless steel vats were filled with 25 L of the must with their skins; 2 vats were used for each PEF treatment applied (4 treatments); and 2 vats were used for untreated samples.

2.2 Pulsed electric fields (PEF) extraction treatments

PEF extraction treatments were exposed in López-Giral et al [3]. The PEF equipment used was an ELCRACK-HVP5 unit (DIL, Germany) with a co-linear PEF treatment chamber ELCRACK DN25 of 2.50 cm of diameter and 2.38 cm distance between electrodes (4.45 cm^2 of electrode area). Pulses of 7.4 kV/cm were applied with frequencies of 300 and 400 Hz and a pulse width of 10 and 20 μs. Denomination of PEF treatments was:

Treat1 (10 μs–300 Hz), Treat2 (10 μs–400 Hz), Treat3 (20 μs–300 Hz), and Treat4 (20 μs–400 Hz).

The crushed grapes from each variety were pumped with a membrane pump (PV8 Saniflo, Wilden, USA) to the PEF treatment chamber.

After treatments, the samples were collected in a stainless steel vat. After 6 h skins and seeds were separated of the must in all samples. Then, the pomace was pressed with a little water press. Aliquots of each sample (control and Treat1–4) were frozen in order to subsequently analyze their aromatic, phenolic, and nitrogen composition.

2.3 Determination of grape volatile compounds by HS-SPME-GC-MS

The grape volatile composition was analyzed according to the methodology exposed by Garde-Cerdán et al. [2]. The SPME fiber used was divinylbenzene/carboxen/polydimethylsiloxane (DVB/CAR/PDMS, 50/30 μm) (Supelco, Bellenfonte, USA). Fibers were thermally conditioned (270°C, 60 min). A total 2 g of NaCl were added to 12 ml of sample into a 20 ml vial for the extraction of volatiles from the different samples. Samples were conditioned for 15 min/60°C with stirring. Subsequently, the extraction was performed for 105 min at this temperature. The SPME fiber containing the volatile compounds was placed in the SPME holder (Supelco) and was manually introduced into the GC injection port at 250 °C (equipped with a glass liner, 0.75 mm I.D. (Supelco), and ThermogreenTM LB-2 septum (Supelco)) and kept during 15 min for desorption. A blank test was carried out to check possible carry-over. The desorbed compounds were separated in an Agilent (Palo Alto, USA) gas chromatograph system (GC) coupled to a mass spectrometric detector (MS) equipped with a SPBTM-20 fused silica capillary column (30 m × 0.25 mm I.D. × 0.25 μm film thickness) (Supelco). Carrier gas used was helium (purity = 99.999%; 1.2 ml/min). The injections were performed in splitless mode (1 min). The program to separate volatile compounds consisted of an initial oven temperature of 40°C for 5 min, a temperature gradient of 2°C/min to a final temperature of 220°C, and a final time of 20 min (total run time = 115 min). The acquisitions were performed in Full Scan (35–300 m/z). NIST library was used for identification by comparison with the mass spectrum and retention index of chromatographic standards. The GC peak area of each compound was obtained from the ion extraction chromatogram by selecting target ions for each one. The analyses were done in duplicate. Hence, the results of volatile compounds correspond to the average of four analyses (n = 4).

2.4 Analysis of grape stilbenes by SPE-HPLC

The determination of these phenolic compounds was performed by the method described by Garde-Cerdán et al. [29]. Briefly, to carry out the extraction of stilbenes from must samples, a Discovery® DCS-18 (100 mg/1 ml) cartridge (Sigma-Aldrich, Madrid, Spain) was employed. The cartridge was conditioned by rinsing with 4 ml of methanol, followed by 4 ml of water. An amount of 10 ml of sample (centrifuged at 10,000 rpm for 10 min) was passed through the solid phase extraction (SPE) cartridge. Then, a washing step was carried out with three fractions of water. Then, the cartridge was dried by letting air pass through it for 30 min. The stilbenes were eluted with 0.7 ml of methanol. The filtered eluate obtained was diluted with water to a proportion of 60:40 (v/v) of methanol/water. The final sample was injected into the HPLC system. Stilbenes were analyzed by reverse-phase HPLC using a liquid chromatograph Agilent 1100 Series. The injected amount was 30 μl and the column temperature was 25 °C. All separations were performed on a ZORBAX Eclipse Plus C18 (150 × 3.0 mm, I.D. 3.5 μm) column (Agilent) with pre-column Eclipse XDB-C18 (12.5 × 4.6 mm, I.D. 5 μm). Three eluents were used as mobile phases: eluent A: water, and acetic acid (98:2, v/v); eluent B: water, acetic acid, and acetonitrile (78:2:20, v/v/v); and eluent C: methanol. The flow rate was 0.9 ml/min. Detection was performed by a fluorescence detector (FLD) and a diode array detector (DAD). The target compounds were identified according to the retention times and UV-Vis spectral characteristics of corresponding standards (Sigma-Aldrich). Quantification was done using the calibration graphs of the respective standards. The SPE-HPLC determinations were carried out in duplicate and, as treatments were performed in duplicate, the results for stilbenes correspond to the mean of four analyses (n = 4).

2.5 Grape amino acids determination by HPLC

The samples amino acids analysis was performed by the method described by Garde-Cerdán et al. [30]. Free amino acids were analyzed using the same HPLC equipment as for stilbenes determination, using both detectors (FLD and DAD). The pure reference compounds and internal standards were obtained from Sigma-Aldrich. Samples were centrifugated at 4.000 rpm/10 min/20 °C. Then, 5 ml of sample was mixed with 100 μl of norvaline and 100 μl of sarcosine (internal standards) and filtered through a 0.45 μm OlimPeak filter (Teknokroma, Barcelona, Spain). Afterward, samples were submitted to automatic precolumn derivatization with o-phthaldialdehyde (OPA Reagent, Agilent) and with 9-fluorenylmethylchloroformate (FMOC Reagent, Agilent). 10 μl at 40°C were injected from the derivatized samples. All separations were made on a Hypersil ODS (250 × 4.0 mm, I.D. 5 μm) column (Agilent).

Eluents used as a mobile phases were: eluent A: 75 mM sodium acetate, 0.018% triethylamine (pH 6.9) + 0.3% tetrahydrofuran; eluent B: water, methanol, and acetonitrile (10:45:45, v/v/v).

Identification of compounds was performed by comparison of their retention times with their pure reference standards. Also the quantification of different amino acids was made by preparing solutions of reference compounds and internal standards in HCl at 0.1 N in the range of the amino acid concentrations usually found in musts Yeast assimilable nitrogen (YAN), was determined according to the method described by Aerny [31]. The results for amino acids and YAN correspond to the mean of four analyses (n = 4) because treatments were performed in duplicate and also analyses of them were carried out in duplicate.

2.6 Statistical analysis

Data management and analysis were performed using SPSS 21.0 (Chicago, USA). ANOVA was used to compare the volatile, phenolic, and nitrogen compounds data.

Results were expressed as means ± standard deviation. A p-value ≤ 0.05 was considered significant (Tukey test). In figures all parameters are listed with their standard deviation. In figures for each grape variety, different letters indicate significant differences. Discriminant analyses were done with the volatile compounds areas and stilbenes and amino acids concentration in the different samples.

3. Results and discussion

3.1 Effect of PEF treatments on volatile composition of Graciano, Tempranillo, and Grenache samples

3.1.1 Monoterpenoids

Monoterpenoids play a significant role in the wine varietal aroma, contributing to its floral and citrus character [12]. The results of monoterpenoids in the control and the samples after each of the four treatments by PEF for the three grape varieties are shown in **Figure 1**. For Graciano, the content of most of monoterpenoids and the total monoterpenoids was higher in control and treatment with the highest energy (Treat4) than in PEF treatments at low energies (Treat1-3). There was an exception in the case of α-terpineol, citronellol, and *p*-cymene, which were not affected by any of the treatments applied to the Graciano variety, except for citronellol in Treat1. In the case of Tempranillo, total monoterpenoids were higher in the treated samples at low energies (Treat1-3). However, when considered individually, these compounds showed no clear trend. In the case of citronellol, nerol, and geraniol, Treat 2 and 3 favored their presence (**Figure 1d**, **e**, and **g**). In the case of linalool, neral, geranyl acetone, and γ-geraniol, no significant differences were observed when comparing control with treated samples. Regarding α-terpineol, its presence in the control was higher than in samples treated with Treat1, 3, and 4 (**Figure 1c**). Moreover, *p*-cymene decreased its amount in the samples with Treat2, 3, and 4 (**Figure 1k**). On the other hand, the presence of total monoterpenoids for Grenache was higher in the samples treated by PEF than in the control one (**Figure 1a**). The same behavior was observed for linalool, citronellol, nerol, geraniol, isogeraniol, and γ -geraniol, with the exception of Treat3 for linalool and Treat2 for isogeraniol, which showed no differences with respect to the control. These results do not match with those obtained by Maza et al. [32], which did not find an increment in the concentration of monoterpenoids on wines of the Grenache variety by application of a PEF treatment. On the other hand, Comuzzo et al. [33] also observed that the concentration of geraniol, significantly, increased after PEF pre-treatment of white grapes of the Garganega variety. Comparing the control must samples with the samples from PEF treatments, Treat1-3 decreased the total monoterpenoids in Graciano by approximately 18% and increased it by about 16% in Tempranillo. PEF increased total monoterpenoids by approximately 50% in Grenache, regardless of the treatment intensity. The overall proportion of monoterpenoids with regard to the total amount of volatile compounds in untreated must samples was of 1.4% in Graciano, 0.9% in Tempranillo, and 4.2% in Grenache. The relative proportion of these compounds in the treated must samples was similar to the control for each of the three grape varieties. Hence it can be said that the effect of PEF technology on monoterpenoids was not selective, as observed by Puértolas et al. [34] and López et al. [35] for phenolic compounds. Monoterpenoids are mainly found in skin, although their distribution between pulp and skin depends on the grape variety and compound [36]. The results obtained suggest

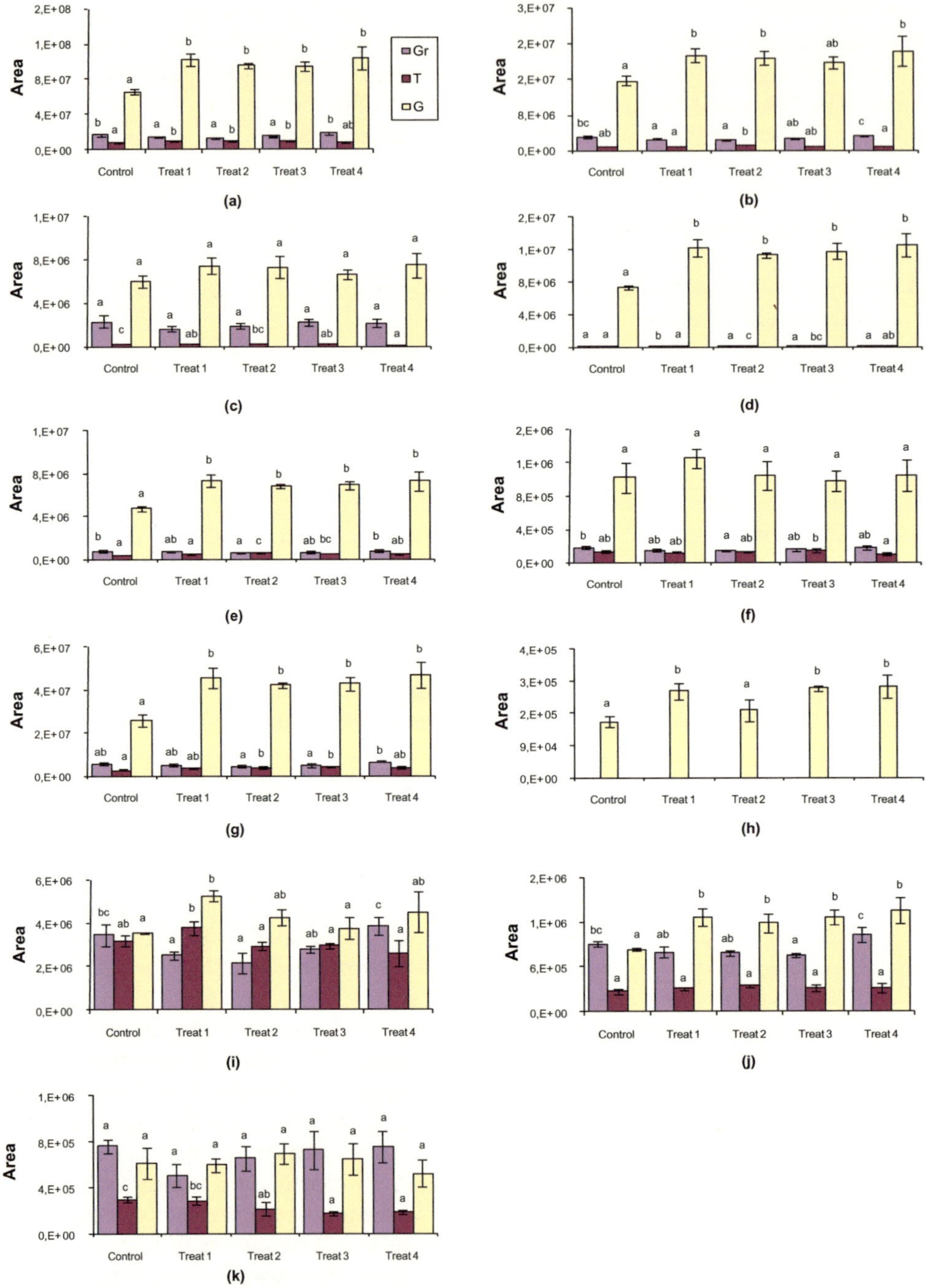

Figure 1.
Monoterpenoids average area in control and PEF treatments 1–4 in Graciano (Gr), Tempranillo (T), and Grenache (G) grape varieties (From Garde-Cerdán et al. [5]).

that monoterpenoids present in the pomace of Grenache were extracted more easily by applying PEF than those present in Tempranillo and Graciano skin because, in this work, PEF application was made in the presence of skin. Other authors have reported that phenolic compounds extraction from grape varieties depends on the morphology and skin composition [37]. Probably, for this reason, a significant increase in monoterpenoids in presence of skin was found in Grenache after PEF treatments. In this regard, Delsart et al. [38] studied two PEF treatments (4 kV/cm; 1 ms and 0.7 kV/cm; 200 ms) on Cabernet Sauvignon grapes. The first PEF treatment, had an impact mainly on vacuolar tannins, whereas the second PEF treatment had a greater impact on the parietal tannins and the skin cell walls. This enhanced polyphenol extraction kinetics. That means the degree of compound extraction from the skin is not only influenced by the grape variety but also by the type of pulse applied.

3.1.2 C_{13} norisoprenoids

Figure 2 shows the results of C_{13} norisoprenoids in the control and the samples after each of the four treatments by PEF for the three grape varieties [5]. In the case of Graciano, the total C_{13} norisoprenoids and (E)-β-damascenone decreased with the lowest energy treatment (Treat1). However, their presence was maintained after treating the samples at higher energy (Treat2–4). The presence of methyl jasmonate decreased by PEF application, except with Treat3 (**Figure 2f**). The (Z)-β-damascenone, β-ionone, and β-cyclocitral contents were not affected by any of the treatments applied to Graciano. For Tempranillo, PEF technology was detrimental to the total content of C_{13} norisoprenoids, and the two isomers of β-damascenone (**Figures 2a–c**). The PEF treatments also decreased the presence of β-ionone upon Treat1 and 4 (**Figure 2d**) but they did not influence the β-cyclocitral and methyl jasmonate amount. For Grenache, only β-ionone and β-cyclocitral were affected by PEF treatments. In general, PEF favored their presence in the musts with the exception of Treat2 and 4 for the β-ionone and Treat2 for the β-cyclocitral (**Figure 2d–e**). These results are in agreement with those obtained by Maza et al. [32], which observed that the concentration of β-ionone, associated with the floral aroma of "violets", which had gone undetected in the control wines, was indeed observed at concentrations greatly exceeding the odor threshold in the wines obtained from Grenache grapes treated by PEF.

Treat1 decreased the presence of total C_{13} norisoprenoids in Graciano by 28%, whereas, in the case of Tempranillo, all the treatments reduced the presence of these compounds by around 37%. However, in Grenache, there was no effect. C_{13} norisoprenoids are distributed in both pulp and skin, unlike monoterpenoids, which are predominantly present in the skin and it could be the cause that no increase was observed with PEF treatments.

In contrast to these results, Comuzzo et al. [11] observed that PEF processing of white grapes (cv. Garganega) after crushing, significantly increased the concentration of norisoprenoid glycosides in the juice of this white grape variety. In control samples, the total concentration of C_{13} norisoprenoids in Graciano, Tempranillo, and Grenache respectively was 7.3%, 2.0%, and 3.2%. The two compounds most abundant in all varieties were β-damascenone (both isomers) and β-ionone. β-Damascenone sum was 97% in Graciano, 92% in Tempranillo, and 98% in Grenache, while the proportion of β-ionone was 1.9% in Graciano, 5.2% in Tempranillo, and 1.3% in Grenache. These proportions varied little upon PEF treatments.

PEF effect was not selective for norisoprenoids. β-Damascenone and β-ionone.

3.1.3 Benzenoid compounds

Figure 3 shows the results for benzenoid compounds in the control and in the samples after each of the four treatments by PEF for the three grape varieties [5].

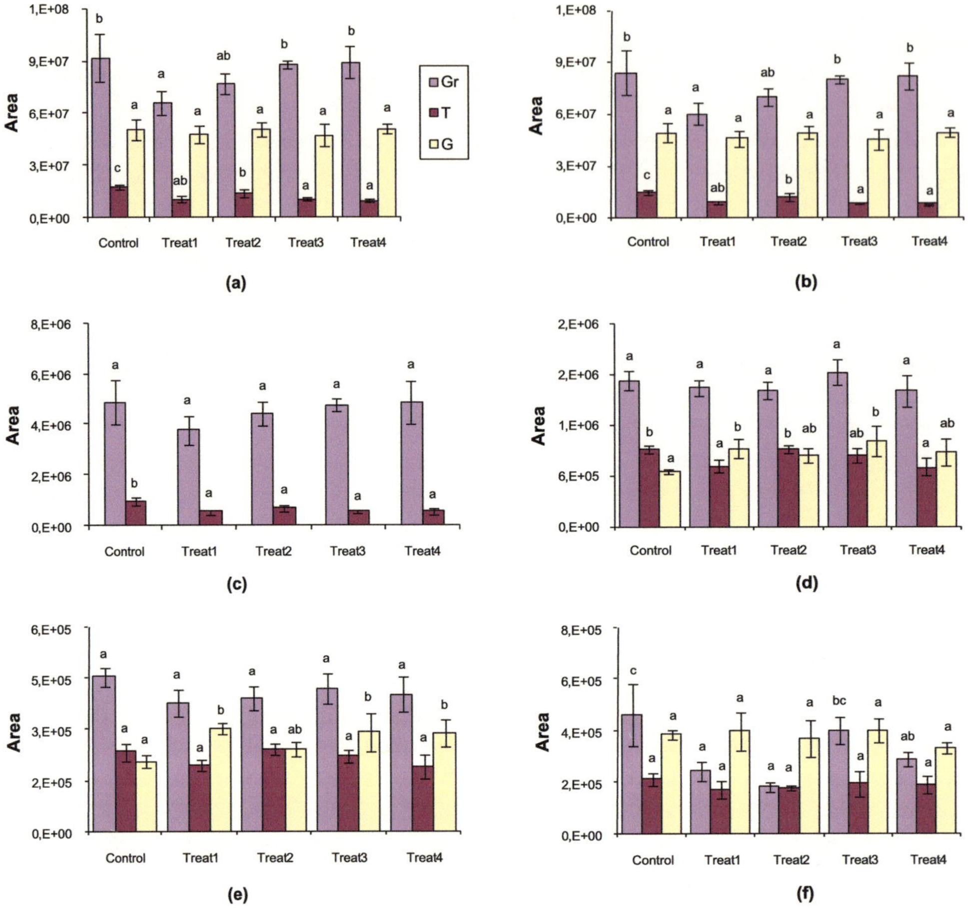

Figure 2.
C_{13} norisoprenoids average area in control and PEF treatments 1–4 in Graciano (Gr), Tempranillo (T), and Grenache (G) grape varieties (From Garde-Cerdán et al. [5]).

Benzenoid compounds, particularly, 2-phenylethanol and eugenol, confer a desirable aroma to the wine, with rose and clove aroma descriptors [17, 39, 40].

In the case of Graciano, treatments had neither effect on total benzenoids nor benzyl alcohol. The content of 2-phenylethanol increased with Treat1, while eugenol was only detected in this grape variety, showing that Treat4 favored its presence. For the Tempranillo variety, Treat1 and 4 resulted in a decrease in the presence of total benzenoids and 2-phenylethanol. On the other hand, benzyl alcohol was found in higher amounts in grapes treated with Treat2. For Grenache, PEF favored the presence of total benzenoids and 2-phenylethanol regardless of the treatment, and Treat1 increased the presence of benzyl alcohol with respect to the control samples. By contrast, Comuzzo et al. [33] observed a slightly decreased of 2-phenylethanol in white wines obtained by PEF processing, but this appears to have a notably low potential impact on sensory perception. By comparing the content of total benzenoids, no loss nor gain was observed upon Graciano samples treated by PEF treatments.

However, in Tempranillo, Treat1 and 4 decreased the presence of these compounds by 24%. For Grenache, all treatments, except Treat2, increased their presence by 45%. The extraction of benzenoid compounds was increased in Grenache by PEF treatments.

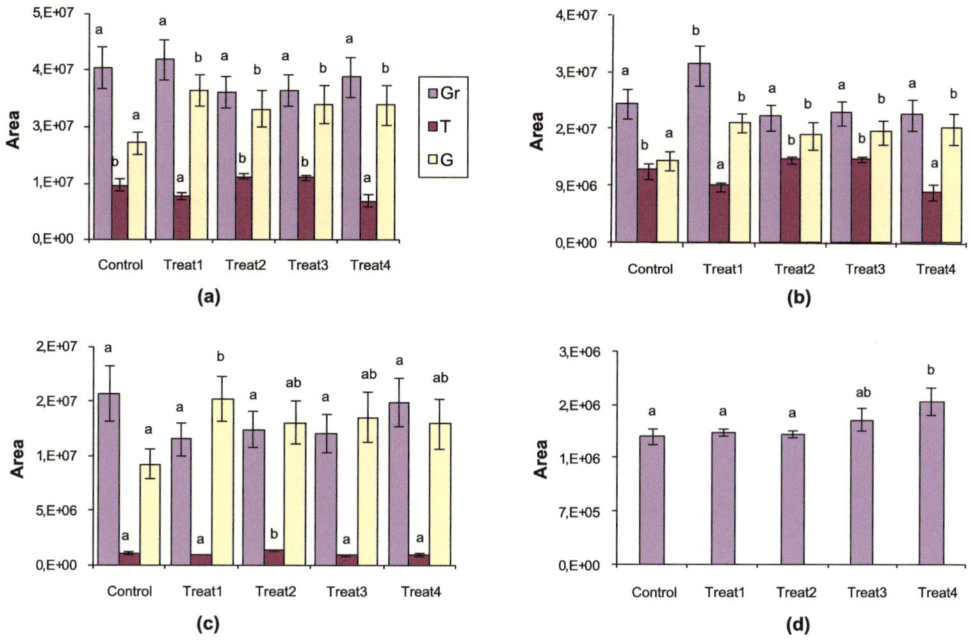

Figure 3.
Benzenoid compounds average area in control and PEF treatments 1–4 in Graciano (Gr), Tempranillo (T), and Grenache (G) grape varieties (From Garde-Cerdán et al. [5]).

3.1.4 Esters

The results of esters in the control and the samples after each of the four treatments by PEF for the three varieties are shown in **Figure 4**. In the case of Graciano, only Treat1 favored the presence of total esters and hexyl acetate. The four PEF treatments resulted in a decrease in 2-hexen-1-ol acetate; and the presence of methyl salicylate was enhanced by Treat2 and 4. For Tempranillo, the treatments applied had no effect on the presence of esters, except for total esters and hexyl acetate in Treat3, which favored its presence; and methyl hexanoate in Treat1-3. For the latter compound, Treat1 decreased its amount, while Treat2 and 3 increased its content in the musts. On the other hand, the presence of total esters for Grenache grape variety was favored by Treat2-4. Moreover, all treatments favored the presence of methyl hexanoate, and an increase for methyl salicylate in the samples was obtained with Treat3 and 4. Maza et al. [32] obtained different results since they did not observe an increment in the concentration of total esters in wines of the Grenache grape variety by application of a PEF treatment. Esters are mainly formed during alcoholic fermentation and play an important role in wine aroma [41]. In control samples, the proportion of total esters with regard to the total amount of volatile compounds was 0.9% in Graciano, 1.1% in Tempranillo, and 0.4% in Grenache. By comparing the contents of these compounds, it can be observed that the application in the Graciano grape variety of Treat1 increased the content of esters by approximately 62%, while Treat3 resulted in a decrease of 23%. In Tempranillo, the application of Treat3 resulted in an increase of these compounds by 29%. Finally, Treat2-4 improved its presence in Grenache by about 32%. Differences depending on grape variety were also found in the study of Fauster et al. [42], where the effects of a PEF treatment on white wine mash were significantly higher for the wines obtained from Traminer variety than those from Grüner Veltliner.

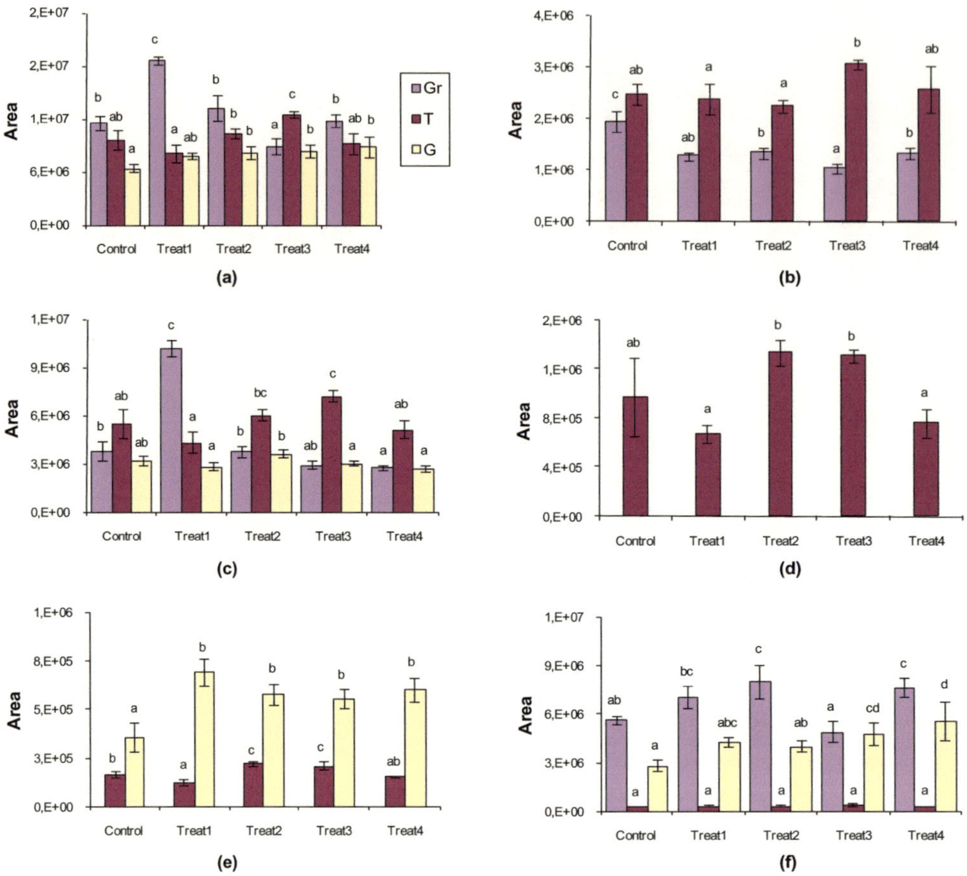

Figure 4.
Esters average area in control and PEF treatments 1–4 in Graciano (Gr), Tempranillo (T), and Grenache (G) grape varieties (From Garde-Cerdán et al. [5]).

3.1.5 C6 compounds

Figure 5 shows the results for C6 compounds in the control and the samples after each of the four treatments by PEF for the three grape varieties [5]. When these compounds are at low levels contribute positively to wine aroma; while, at high levels, they are responsible for herbaceous flavors [33].

The presence of (E)-2-hexen-1-ol in must samples of Treat1 of Graciano was diminished. However, Treat2 in Tempranillo enhanced the content of total C6 compounds, n-hexanol, (Z)-3-hexen-1-ol, and (E)-2-hexenal; while Treat1 decreased the content of hexanal. Treat1 in Grenache favored the presence of total C6 compounds and (Z)-3-hexen-1-ol. Also Treat1 and 3 enhanced the content of hexanal, and Treat2 favored the presence of (E)-2-hexenal. In general, and matching with the Comuzzo et al. [33] results, the PEF treatments hardly affected the amounts of C6 compounds. The PEF samples increased of total C6 compounds in 72% after Treat1 in Grenache and 31% upon Treat2 in Tempranillo. On the other hand, the most abundant C6 compounds were in control samples accounting for 87%, 95%, and 91% in Graciano, Tempranillo, and Grenache respectively. The PEF treatments did not affect the relative abundance of C6 compounds in must samples.

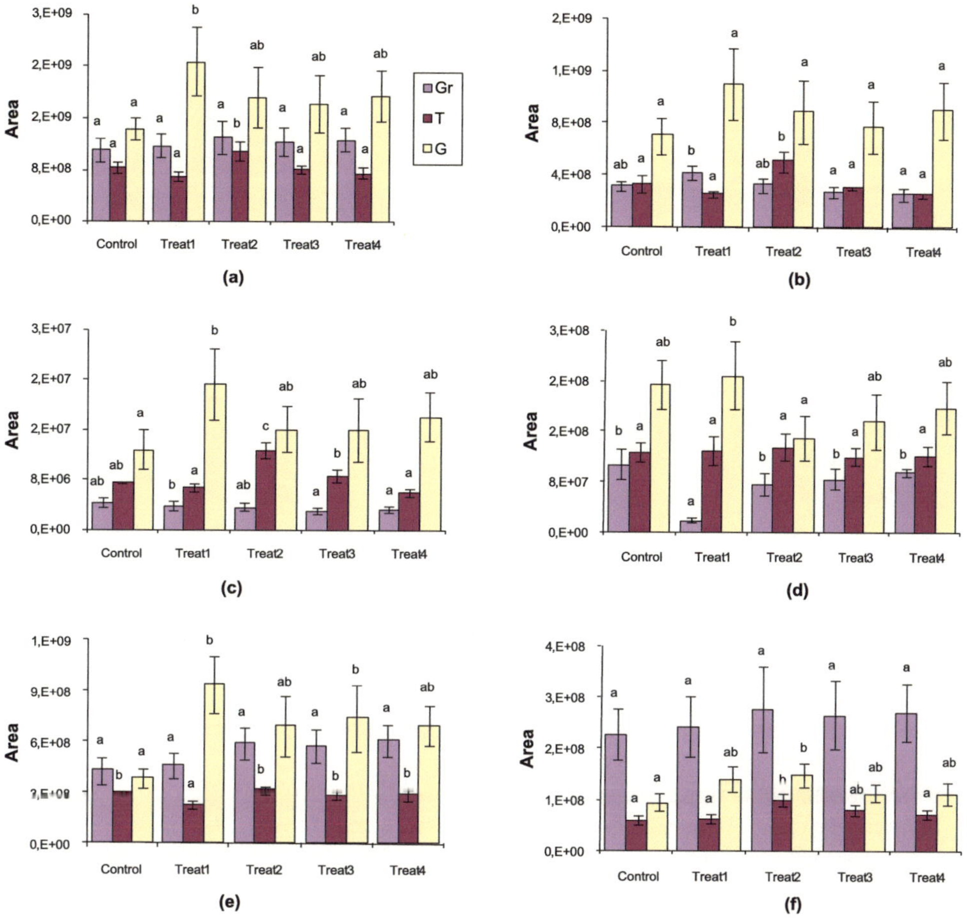

Figure 5.
C6 compounds average area in control and PEF treatments 1–4 in Graciano (Gr), Tempranillo (T), and Grenache (G) grape varieties (From Garde-Cerdán et al. [5]).

3.2 Influence of PEF treatments on resveratrol and piceid content in Graciano, Tempranillo, and Grenache samples

The concentration of total stilbenes, trans-resveratrol, *trans-* and *cis*-piceid from each of the three grape varieties in the control and the four PEF treatments tested were shown in **Figure 6**. Among the three grape varieties treated, Graciano presented the highest total stilbenes concentration (**Figure 6a**), is the only one in which *trans*-resveratrol was detected (**Figure 6b**). For its part, *cis*-resveratrol was not detected in any of the samples because it is usually absent in grapes [43]. The total stilbene content in the PEF treated Tempranillo samples increased more (up to 200%, ranging from 90% to 200%) with respect to the untreated samples that both, the Graciano (40% for Treat2 and Treat3, and 60% for Treat4) and Grenache (50% with the Treat4) samples. Thus, the Treat4, the one with the highest time and energy, was the most effective on the total stilbene extraction for the three grape varieties (**Figure 6a**). Therefore, as the phenolic compounds are found mainly in the grape skins, the application of PEF treatments increased their extraction although the efficiency and the effect of this PEF on the stilbenes depend on the grape variety applied, which plays

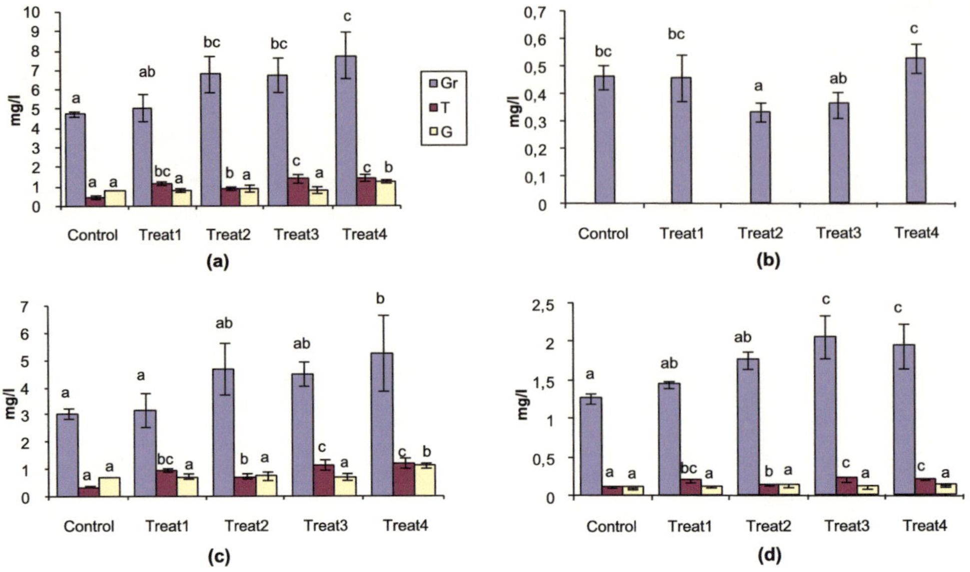

Figure 6.
Stilbene concentration (mg/l) in control and PEF treatments 1–4 in Graciano (Gr), Tempranillo (T), and Grenache (G) grape varieties (From López-Alfaro et al. [4]).

an important role in the grapes stilbene concentration [44]. The piceids, which are extracted from the grape skins before extracting the aglycones, resveratrols since the presence of ethanol is required to increase their solubility and mobility [45], were in higher concentration (**Figure 6c** and **d**) than resveratrol (**Figure 6b**). This matched with those results reported by Romero-Pérez et al. [46]. Besides, trans-piceid was the most abundant stilbene in the three cultivars, improving its extraction in Tempranillo with all the PEF applied treatments, while in the case of the Graciano and Grenache grape varieties, increases its content only after applying the highest energy and timing treatment (Treat4) (**Figure 6c**). *trans*-Piceid is the precursor of *trans*-resveratrol, in which content did not increase in Graciano grapes with the application of PEF treatments (**Figure 6b**). On the other hand, Treat3 and Treat4 increased the *cis*-piceid extraction in Graciano (**Figure 6d**), meanwhile, as observed with the *trans*-isomer for the Tempranillo grape variety, the *cis*-piceid extraction increased with all the PEF treatments for this grape variety. Regarding the effects of the PEF on Grenache *cis*-piceid content, no one change was observed (**Figure 6d**). Similar as what occurs with the *trans* isomer, *cis*-piceid is the precursor of *cis*-resveratrol by hydrolysis during fermentation. Besides, the final concentration of these compounds is conditioned by the reaction of *trans-cis* isomerization, especially in relation to the aglycons, due to the instability of the *cis* isomer, so the *trans* is the highest and most stable isomer [17].

3.3 Effect of PEF treatments on vnitrogen compounds of Graciano, Tempranillo, and Grenache samples

The content of total amino acids, total amino acids without proline, and YAN from each of the three grape varieties in the control and the four PEF treatments tested were shown in **Figure 7**. The effect of PEF treatments in the extraction of nitrogen compounds was different between the three grape varieties. Control samples of

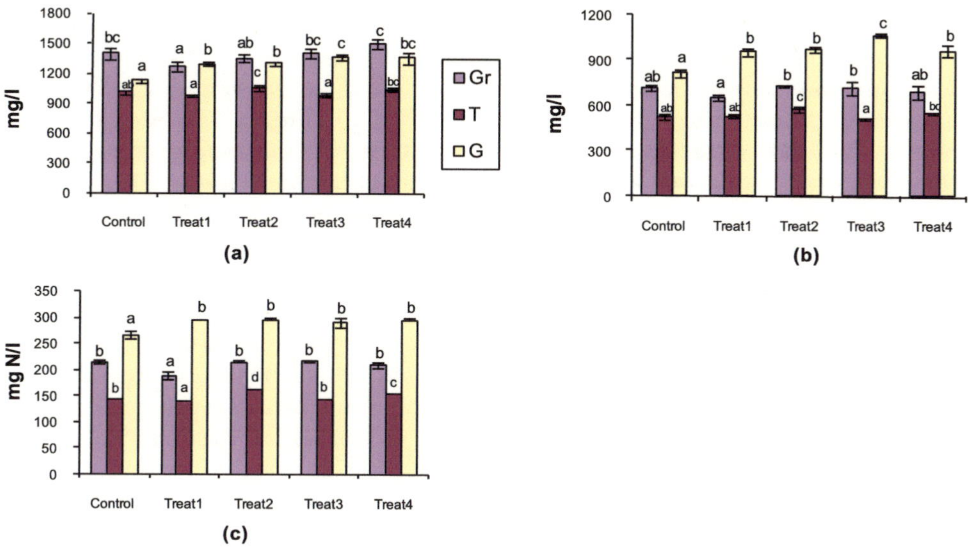

Figure 7.
Total amino acids (mg/l), total amino acids without proline (mg/l) and yeast assimilable nitrogen (YAN) (mg N/l) in Graciano (Gr), Tempranillo (T), and Grenache (G) grapes for control and PEF treatments (Treats 1–4).

Graciano only presented significant differences of total amino acids with samples of Treat1 (lower than control) (**Figure 7a**). However, in Tempranillo, samples of control, Treat1 and 3 showed no significant differences between them, with higher amino acids content only in samples treated with Treat2. In the Grenache grape variety, the total amino acids content in all PEF samples was significantly different and higher than the control ones. Samples with the highest total amino acids content were those treated with Treat3 and 4 for this grape variety. In the case of the total amino acids without proline (**Figure 7b**), no differences were observed between control and PEF treated samples in Graciano. In Tempranillo, only samples of Treat2 showed positive differences with regard to control ones. Meanwhile, in Grenache, all the PEF treated samples obtained significantly higher values of total amino acids without proline than the control, being Treat3 the best treatment, matching with that observed for total amino acids content. The ratio of total amino acids without proline with regard to total amino acids (in %) was dependent on the grape variety. Grenache was the variety most benefited by the application of PEF treatments, being all the PEF treatments applied significantly advantageous with respect to not applying them. The percentage of improvement of the Grenache samples treated by PEF with respect to the control samples ranged between 15% and 22% for the total amino acids and from 16 to 33% for the total amino acids without proline. Regarding the effect in of PEF treatments on YAN content, in Graciano the YAN content was around 200 mg N/l, being the samples of Treat1 the ones with significantly lower YAN content (**Figure 7c**). In Tempranilllo, all samples presented YAN values between 140 and 162 mg N/l, being only in the samples from Treat2 and 4 superiors to 150 mg N/l, which is the minimum value to achieve the correct development of the fermentation, according to Blouin and Peynaud [47] and Bell and Henschke [48]. Tempranillo samples treated with Treat1 presented the lowest YAN content and it was significantly different from the rest of the treatments. In Grenache, samples treated by PEF did not show any differences in the YAN content between them with values around 292 mg

N/l. However, Grenache control samples had less YAN content (266 mg N/l) than the PEF treated samples. YAN is composed of ammonium ions and free amino nitrogen. These compounds are the main sources of nitrogen for *Saccharomyces cerevisiae* and their content affects the kinetics of fermentation [49, 50]. **Figure 8** shows the free amino acids found in Graciano, Tempranillo, and Grenache musts from control and PEF treatments (Treats 1-4). Proline (**Figure 8t**) and arginine (**Figure 8i**) were the two majority amino acids in grape musts. Graciano was the grape variety with more proline, with contents of about 570–840 mg/l, and Grenache was the variety with less proline, with contents of about 280–420 mg/l. With respect to the arginine content, Grenache had more content than Tempranillo and Graciano, with values about 190–315 mg/l, 120–155 mg/l, and 140–230 mg/l, respectively. The characteristic index based on the ratio of the proline and arginine is used to indicate the differential accumulation of these amino acids by different grape cultivars. The proportion of non-assimilable (proline) to assimilable nitrogen (arginine) is indicated by this index, which provides a useful indication of the likely nutritional value for yeast metabolism of the grape must [27]. Regarding the effects of the PEF treatments on the proline and arginine content, in Graciano, there were significant differences of proline content between samples of Treat4, which had the highest values, and the rest of the treatments. For arginine, the highest content was observed for the samples treated by PEF with the Treat2, and the lowest content was for Treat4 samples. In Tempranillo, samples of Treat1 had significantly less content of proline than the rest of the samples, except Treat3, among which there were no differences. On the other hand, all Tempranillo PEF treated samples were richer in arginine content than the control one, being those of Treat1 and 4, the ones with more content of this amino acid. In Grenache, proline content of Treat1 and 4 was higher than control samples and for arginine concentration, content in Treat2 and 3 were higher with respect to control ones. The highest arginine/proline ratio was observed in the Grenache variety with values between 0.57 and 0.98. The highest value of this ratio was for Treat3 samples. Values of this ratio for Graciano were between 0.19 and 0.35 and the best ratio was for Treat2 samples. Values of this ratio for Tempranillo were between 0.27 and 0.34 and the best ratio was for Treat1 samples. Thus, as the proline to arginine ratio was lower than 1 for the three grape varieties, that suggested that all red varieties are proline accumulators, as reported by authors as Garde-Cerdán et al. [50], and Pérez-Álvarez et al. [51] for cv. Tempranillo, and that the proline to arginine ratio is influenced by nitrogen nutrition [52] and depends on the variety [48]. Amino acids concentration at harvest depends on climatic conditions and agronomic practices, while amino acid profile mainly depends on variety and zone [53, 54] and also the maturity. For example, Grenache at an early stage of maturity (19.7 °Brix) was classified as a high arginine accumulator, but by 24 °Brix it accumulated predominantly more proline than arginine [55]. In this study, Grenache with 24 °Brix, also presented more content of proline than arginine. Leaving aside the proline and arginine which have already been said were the majority amino acids in samples, in Graciano, the four amino acids with more content, from highest to lowest content, were alanine, glutamic acid, serine, and threonine, all of them with contents from 60 mg/l to 156 mg/l. In Tempranillo, glutamic acid, alanine, serine and histidine, with contents between 35 and 85 mg/l, and in Grenache, glutamic acid, histidine, alanine, and threonine, with contents from 48 to 172 mg/l were the amino acids majority. Among the sources that produce rapid yeast growth are amino acids such as glutamine, asparagine, glutamic acid, and alanine. These produce carbon derivatives, which are rapidly integrated into the fermentative metabolism of yeasts. In this study, except for asparagine, whose

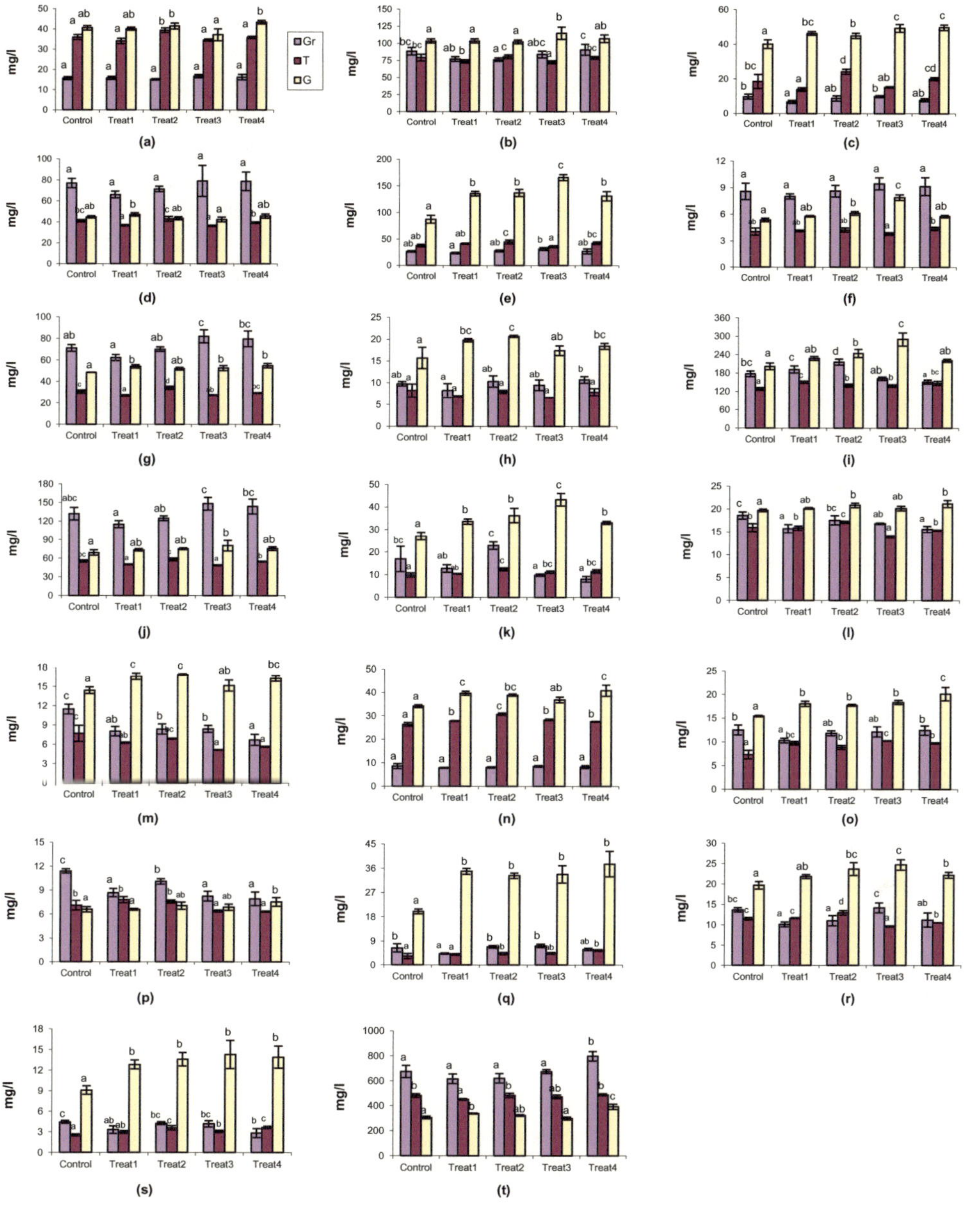

Figure 8.
Free amino acids concentration (mg/l) in Graciano (Gr), Tempranillo (T), and Grenache (G) grapes for control and PEF treatments (Treats 1–4).

concentrations were low (except in Grenache, with a content ranging between 40 and 50 mg/l), both, alanine and glutamic acid were between the first and third place of the amino acids in the three grape varieties (without taking into account proline and arginine). On the other hand, among the sources that produce lower growth are the aromatic amino acids (tryptophan, tyrosine, and phenylalanine) and branched (leucine, isoleucine, and valine) that produce ketoacid complexes, which must be converted into aldehyde complexes and higher alcohols for their elimination [56]. In the case of these latter groups of amino acids, the fact of producing carbon skeletons that are not easily assimilated also makes these amino acids the most interesting in the production of aromas [57]. In this study, the concentration of these amino acids was low for the three grape varieties: so, their content was lower than 20 mg/l, 31 mg/l, and 40 mg/l in Graciano, Tempranillo, and Grenache must samples, respectively. Threonine, methionine, and serine are also considered aroma precursors. The concentrations of some of the volatiles correlated well with the aromatic composition of the equivalent wines. Development of models by chemometric analysis showed that threonine and serine affected corresponding fatty acid esters and alcohols, and methionine strongly affected methionol concentration [48]. In this study, threonine content in Graciano ranged between 60 and 85 mg/l, in Grenache around 50 mg/l, and Tempranillo between 25 and 35 mg/l. For methionine, the variety with the highest content was Grenache with values around 15 mg/l. Serine content in Graciano, Tempranillo, and Grenache was higher than 60, 35, and 40 mg/l, respectively. These amino acids are very important because higher alcohols come from them directly and some esters indirectly, as they come from these higher alcohols. For example, n-propanol comes from threonine, 2-methyl-1-butanol from isoleucine, 3-methyl-1-butanol from leucine, isobutanol from valine, 2-phenylethanol from phenylalanine, and methionol from methionine [58]. With respect to the effect of the four PEF treatments on the content of amino acids in the musts of the three grape varieties, the results were very dependent on the grape variety. In PEF treated samples of Graciano, only threonine content (**Figure 8g**) in Treat3 samples was superior to control ones. In Tempranillo, the amino acids contents of the Treat2 samples were higher than the control one (for aspartic acid, asparagine histidine, threonine, tyrosine, valine, leucine, phenylalanine, and lysine). No differences or even reduction of some amino acids contents were produced by Treat1, 3, and 4. In some cases, positive differences were obtained with Treat4 respect to the control. In Grenache, in general, all the samples treated by PEF showed higher or equal amino acids contents than the control ones (**Figure 8**).

4. Conclusions

The technology of pulsed electric fields (PEF) affected the volatile composition of Graciano, Tempranillo, and Grenache depending on the grape variety. However, the flavor profile of the samples was not affected in any case. The volatile composition of grape juice was enhanced by PEF application in Grenache, without significant improvement in Graciano and Tempranillo. However, the highest energy PEF treatment improved the stilbene total content in musts from the three grape varieties in an important way. Moreover, the four PEF treatments enhanced YAN and the total amino acids content in Grenache, while for Graciano and Tempranillo grape varieties, the PEF treatments hardly influenced the grape nitrogen composition. In conclusion, PEF is a good tool in order to improve the quality of the Grenache grape variety.

Acknowledgements

This research was funded by the Instituto Nacional de Investigación y Tecnología Agraria y Alimentaria (INIA) to project RTA201100070-00-00 and the Navarra Government to project IIQ14037.RI1.

Author details

Teresa Garde-Cerdán[1*], Eva P. Pérez-Álvarez[1], Pilar Rubio-Bretón[1] and Noelia López-Giral[2*]

1 Grupo VIENAP, Instituto de Ciencias de la Vid y del Vino (CSIC, Gobierno de La Rioja, Universidad de La Rioja), Logroño, La Rioja, Spain

2 National Centre for Food Technology and Safety (CNTA), San Adrián, Navarra, Spain

*Address all correspondence to: teresa.garde.cerdan@csic.es and nlopez@cnta.es

References

[1] Garde-Cerdán T, Arias-Gil M, Marsellés-Fontanet AR, Ancín-Azpilicueta C, Martín-Belloso O. Effects of thermal and non-thermal processing treatments on fatty acids and free amino acids of grape juice. Food Control. 2007;**18**:473-479. DOI: 10.1016/j.foodcont.2005.12.004

[2] Donsì F, Ferrari G, Pataro G. Applications of pulsed electric field treatments for the enhancement of mass transfer from vegetable tissue. Food Engineering Reviews. 2010;**2**:109-130. DOI: 10.1007/s12393-010-9015-3

[3] López-Giral N, González-Arenzana L, González-Ferrero C, López R, Santamaría P, López-Alfaro I, et al. Pulsed electric field treatment to improve the phenolic compound extraction from Graciano, Tempranillo and Grenache grape varieties during two vintages. Innovative Food Science and Emerging Technologies. 2015;**28**:31-39. DOI: 10.1016/j.ifset.2015.01.003

[4] López-Alfaro I, González-Arenzana L, López N, Santamaría P, López R, Garde-Cerdán T. Pulsed electric field treatment enhanced stilbene content in Graciano, Tempranillo and Grenache grape varieties. Food Chemistry. 2013;**141**:3759-3765. DOI: 10.1016/j.foodchem.2013.06.082

[5] Garde-Cerdán T, González-Arenzana L, López N, López R, Santamaría P, López-Alfaro I. Effect of different pulsed electric field treatments on the volatile composition of Graciano, Tempranillo and Grenache grape varieties. Innovative Food Science and Emerging Technologies. 2013;**20**:91-99. DOI: 10.1016/j.ifset.2013.08.008

[6] Rubio-Bretón P, Salinas MR, Nevares I, Pérez-Álvarez EP, del Álamo-Sanza M-SRS, Alonso GL, et al. Recent advances in the study of grape and wine volatile composition: Varietal, fermentative, and aging aroma compounds. In: Bordiga M, Nollet LML, editors. Food Aroma Evolution. During Food Processing, Cooking, and Aging. 1st ed. Boca Raton: CRC Press; 2020. pp. 439-464

[7] Cabrita MJ, Costa Freitas AM, Laureano O, Borsa D, Di Stefano R. Aroma compounds in varietal wines from Alentejo, Portugal. Journal of Food Compounds Analytical. 2007;**20**:375-390. DOI: 10.1016/j.jfca.2006.12.006

[8] Perestrelo R, Barros AS, Rocha SM, Câmara JS. Optimisation of solid-phase microextraction combined with gas chromatography-mass spectrometry based methodology to establish the global volatile signature in pulp and skin of *Vitis vinifera* L. grape varieties. Talanta. 2011;**85**:1483-1493. DOI: 10.1016/j.talanta.2011.06.025

[9] Zalacain A, Marín J, Alonso GL, Salinas MR. Analysis of wine primary aroma compounds by stir bar sorptive extraction. Talanta. 2007;**71**:1610-1615. DOI: 10.1016/j.talanta.2006.07.051

[10] Moreno-Arribas MV, Polo MC. Wine Chemistry and Biochemistry. New York: Springer; 2009. p. 752. DOI: 10.1007/978-0-387-74118-5

[11] Garde-Cerdán T, Santamaría P, Rubio-Bretón P, González-Arenzana L, López-Alfaro I, López R. Foliar application of proline, phenylalanine, and urea to Tempranillo vines: Effect on grape volatile composition and comparison with the use of commercial nitrogen fertilizers. LWT- Food Science and Technology. 2015;**60**:684-689. DOI: 10.1016/j.lwt.2014.10.028

[12] Ribéreau-Gayon P, Glories Y, Maujean A, Dubourdieu D. Handbook of Enology: The Chemistry of Wine Stabilization and Treatments. 2nd ed. Chichester: John Wiley & Sons; 2006. p. 450

[13] Pedroza MA, Zalacain A, Lara JF, Salinas MR. Global grape aroma potential and its individual analysis by SBSE-GC-MS. Food Research International. 2010;**43**:1003-1008. DOI: 10.1016/j.foodres.2010.01.008

[14] Fernández-Mar MI, Mateos R, García-Parrilla MC, Puertas B, Cantos-Villar E. Bioactive compounds in wine: Resveratrol, hydroxytyrosol and melatonin: A review. Food Chemistry. 2012;**130**:797-813. DOI: 10.1016/j.foodchem.2011.08.023

[15] Gambini J, López-Grueso R, Olaso-González G, Inglés M, Abdelazid K, El Alami M, et al. Resveratrol: Distribución, propiedades y perspectivas. Revista Española de Geriatría y Gerontología. 2013;**48**:79-88. DOI: 10.1016/j.regg.2012.04.007

[16] Portu J, Santamaría P, López-Alfaro I, López R, Garde-Cerdán T. Methyl jasmonate foliar application to Tempranillo vineyard improved grape and wine phenolic content. Journal of Agricultural and Food Chemistry. 2015;**63**:2328-2337. DOI: 10.1021/jf5060672

[17] Garde-Cerdán T, Portu J, López R, Santamaría P. Effect of foliar applications of proline, phenylalanine, urea, and commercial nitrogen fertilizers on stilbene concentrations in Tempranillo musts and wines. American Journal of Enology and Viticulture. 2015;**66**:542-547. DOI: 10.5344/ajev.2015.14128

[18] Sablayrolles JM, Dubois C, Manginot C, Roustan JL, Barre P. Effectiveness of combined ammoniacal nitrogen and oxygen additions for completion of sluggish and stuck fermentations. Journal of Fermentation and Bioengineering. 1996;**82**:377-381. DOI: 10.1016/0922-338X(96)89154-9

[19] Arias-Gil M, Garde-Cerdán T, Ancín-Azpilicueta C. Influence of addition of ammonium and different amino acid concentrations on nitrogen metabolism in spontaneous must fermentation. Food Chemistry. 2007;**103**:1312-1318. DOI: 10.1016/j.foodchem.2006.10.037

[20] Garde-Cerdán T, Martínez-Gil AM, Lorenzo C, Lara JF, Pardo F, Salinas MR. Implications of nitrogen compounds during alcoholic fermentation from some grape varieties at different maturation stages and cultivation systems. Food Chemistry. 2011;**124**:106-116. DOI: 10.1016/j.foodchem.2010.05.112

[21] Fairbairn S, McKinnon A, Musarurwa HT, Ferreira AC, Bauer FF. The impact of single amino acids on growth and volatile aroma production by *Saccharomyces cerevisiae* strains. Frontiers in Microbiology. 2017;**8**:2554. DOI: 10.3389/fmicb.2017.02554

[22] Gobert A, Tourdot-Marécha R, Sparrow C, Morge C, Alexandre H. Influence of nitrogen status in wine alcoholic fermentation. Food Microbiology. 2019;**83**:71-85. DOI: 10.1016/j.fm.2019.04.008

[23] Pérez-Álvarez EP, Martínez-Vidaurre JM, García-Escudero E, Garde-Cerdán T. Amino acids content in 'Tempranillo' must from three soil types over four vintages. Vitis Journal of Grapevine Research. 2019;**58**:3-12. DOI: 10.5073/vitis.2019.58.special-issue.3-12

[24] Francis IL, Newton JL. Determining wine aroma from compositional data.

Australian Journal of Grape and Wine Research. 2005;**11**:114-126. DOI: 10.1111/j.1755-0238.2005.tb00283.x

[25] Swiegers JH, Bartowsky EJ, Henschke PA, Pretorius IS. Yeast and bacterial modulation of wine aroma and flavour. Australian Journal of Grape and Wine Research. 2005;**11**:139-173. tDOI: 10.1111/j.1755-0238.2005.tb00285.x

[26] Garde-Cerdán T, Ancín-Azpilicueta C. Effect of the addition of different quantities of amino acids to nitrogen-deficient must of the formation of esters alcohols, and acids during wine alcoholic fermentation. LWT- Food Science and Technology. 2008;**41**:501-510. DOI: 10.1016/j.lwt.2007.03.018

[27] Martínez-Gil AM, Garde-Cerdán T, Lorenzo C, Lara JF, Pardo F, Salinas MR. Volatile compounds formation in alcoholic fermentation from grapes collected at 2 maturation stages: Influence of nitrogen compounds and grape variety. Journal of Food Science. 2012;77:C71-C79. DOI: 10.1111/j.1750-3841.2011.02441.x

[28] Malherbe S, Watts V, Nieuwoudt HH, Bauer FF, Toit MDU. Analysis of volatile profiles of fermenting grape must by headspace solid-phase dynamic extraction coupled with gas chromatography-mass spectrometry (HS-SPDE GC-MS): Novel application to investigate problem fermentations. Journal of Agricultural and Food Chemistry. 2009;**57**:5161-5166. DOI: 10.1021/jf900532v

[29] Garde-Cerdán T, Gutiérrez-Gamboa G, Baroja E, Rubio-Bretón P, Pérez-Álvarez EP. Influence of methyl jasmonate foliar application to vineyard on grape volatile composition over three consecutive vintages. Food Research International. 2018;**112**:274-283. DOI: 10.1016/j.foodres.2018.06.04

[30] Garde-Cerdán T, Portu J, López R, Santamaría P. Effect of methyl jasmonate application to grapevine leaves on grape amino acid content. Food Chemistry. 2016;**203**:536-539. DOI: 10.1016/j.foodchem.2016.02.049

[31] Aerny J. Compostes azotes des mouts et des vins. Revue Suisse de Viticulture, d'Arboriculture et d'Horticulture. 1996;**28**:161-165

[32] Maza MA, Martínez JM, Hernández-Orte P, Cebrián G, Sánchez-Gimeno AC, Álvarez I, et al. Influence of pulsed electric fields on aroma and polyphenolic compounds of Garnacha wine. Food and Bioproducts Processing. 2019;**116**:249-257. DOI: 10.1016/j.fbp.2019.06.005

[33] Comuzzo P, Marconi M, Zanella G, Querzè M. Pulsed electric field processing of white grapes (cv. Garganega): Effects on wine composition and volatile compounds. Food Chemistry. 2018;**264**:16-23. DOI: 10.1016/j.foodchem.2018.04.116

[34] Puértolas E, Hernández-Orte P, Sladaña G, Álvarez I, Raso J. Improvement of winemaking process using pulsed electric fields at pilot-plant scale. Evolution of chromatic parameters and phenolic content of Cabernet Sauvignon red wines. Food Research International. 2010;**43**:761-766. DOI: 10.1016/j.foodres.2009.11.005

[35] López N, Puértolas E, Hernández-Orte P, Álvarez I, Raso J. Effect of a pulsed electric field treatment on the anthocyanins composition and other quality parameters of Cabernet Sauvignon freshly fermented model wines obtained after different maceration times. Lebensmittel-Wissenschaft und Technologie. 2009;**42**:1225-1231. DOI: 10.1016/j.lwt.2009.03.009

[36] Mateo JJ, Jiménez M. Monoterpenes in grape juice and wines. Journal of Chromatography A. 2000;**881**:557-567. DOI: 10.1016/s0021-9673(99)01342-4

[37] Puértolas E, López N, Condón S, Álvarez I, Raso J. Potential applications of PEF to improve red wine quality. Trends in Food Science and Technology. 2010;**21**:247-255. DOI: 10.1016/j.tifs.2010.02.002

[38] Delsart C, Cholet C, Ghidossi R, Grimi N, Gontier E, Gény L, et al. Effects of pulsed electric fields on Cabernet Sauvignon grape berries and on the characteristics of wines. Food and Bioprocess Technology. 2014;7:424-436. DOI: 10.1007/s11947-012-1039-7

[39] Garde-Cerdán T, Rubio-Bretón P, Marín-San Román S, Sáenz de Urturi I, Pérez-Álvarez EP. Pre-fermentative maceration with SO_2 enhanced the must aromatic composition. Food Chemistry. 2021;**345**:128870

[40] Rubio-Bretón P, Lorenzo C, Salinas MR, Martínez J, Garde-Cerdán T. Influence of oak barrel aging on the quality of red wines. In: Grão CA, Chuteira AB, editors. Oak: Ecology, Types and Management. New York: Nova Science Publishers; 2012. pp. 59-86

[41] Jackson RS. Wine Science: Principles and Applications. New York: Academic Press; 2008. p. 751

[42] Fauster T, Philipp C, Hanz K, Scheibelberger R, Teufl T, Nauer S, et al. Impact of a combined pulsed electric field (PEF) and enzymatic mash treatment on yield, fermentation behaviour and composition of white wine. European Food Research and Technology. 2020;**246**:609-620. DOI: 10.1007/s00217-020-03427-w

[43] Gatto P, Vrhovsek U, Muth J, Segala C, Romualdi C, Fontana P, et al. Ripening and genotype control stilbene accumulation in healthy grapes. Journal of Agricultural and Food Chemistry. 2008;**56**:11773-11785. DOI: 10.1021/jf8017707

[44] Bavaresco L, Mattivi F, De Rosso M, Flamini R. Effects of elicitors, viticultural factors, and enological practices on resveratrol and stilbenes in grapevine and wine. Mini Reviews in Medicinal Chemistry. 2012;**12**:1366-1381. DOI: 10.2174/13895575112091366

[45] Mattivi F, Reniero F, Korhammer S. Isolation, characterization, and evolution in red wine vinification of resveratrol monomers. Journal of Agricultural and Food Chemistry. 1995;**43**:1820-1823. DOI: 10.1021/jf00055a013

[46] Romero-Pérez AI, Ibern-Gómez M, Lamuela-Raventós RM, de La Torre-Boronat MC. Piceid, the major resveratrol derivative in grape juices. Journal of Agricultural and Food Chemistry. 1999;**47**:1533-1536. DOI: 10.1021/jf981024g

[47] Blouin J, Peynaud E. Enología Práctica: Conocimiento y Elaboración del Vino. Madrid: Mundi-Prensa; 2003. p. 360

[48] Bell S-J, Henschke PA. Implications of nitrogen nutrition for grapes, fermentation and wine. Australian Journal of Grape and Wine Research. 2005;**11**:242-295. DOI: 10.1111/j.1755-0238.2005.tb00028.x

[49] Gardner JM, Poole K, Jiranek V. Practical significance of relative assimilable nitrogen requirements of yeast: A preliminary study of fermentation performance and liberation of H_2S. Australian Journal of Grape and Wine Research.

2002;**8**:175-179. DOI: 10.1111/j.1755-0238.2002.tb00253.x

[50] Garde-Cerdan T, Gutiérrez-Gamboa G, López R, Rubio-Bretón P, Pérez-Álvarez EP. Influence of foliar application of phenylalanine and urea at two doses to vineyards on grape volatile composition and amino acids content. 2018;**57**:137-141. DOI: 10.5073/VITIS.2018.57.137-141

[51] Pérez-Álvarez EP, Garde-Cerdán T, García-Escudero E, Martínez-Vidaurre JM. Effect of two doses of urea foliar application on leaves and grape nitrogen composition during two vintages: Effect of urea foliar applications on grape nitrogen compounds. Journal of the Science of Food and Agriculture. 2017;**97**:2524-2532. DOI: 10.1002/jsfa.8069

[52] Hilbert G, Soyer JP, Molot C, Giraudon J, Milin S, Gaudillere JP. Effects of nitrogen supply on must quality and anthocyanin accumulation in berries of cv. Merlot Vitis. 2003;**42**:69-76

[53] Huang Z, Ough CS. Amino acid profiles of commercial grape juices and wines. American Journal of Enology and Viticulture. 1991;**42**:261-267

[54] Garde-Cerdán T, Lorenzo C, Lara JF, Pardo F, Ancín-Azpilicueta C, Salinas MR. Study of the evolution of nitrogen compounds during grape ripening. Application to differentiate grape varieties and cultivated systems. Journal of Agricultural Food Chemistry. 2009;**57**:2410-2419. DOI: 10.1021/jf8037049

[55] Kliewer WM. Free amino acids and other nitrogenous fractions in wine grapes. Journal of Food Science. 1970;**35**:17-21. DOI: 10.1111/j.1365-2621.1970.tb12358.x

[56] Beltran G, Esteve-Zarzoso B, Rozès N, Mas A, Guillamón JM. Influence of the timing of nitrogen additions during synthetic grape must fermentations on fermentation kinetics and nitrogen consumption. Journal of Agricultural and Food Chemistry. 2005;**53**:996-1002. DOI: 10.1021/jf0487001

[57] González B, Vázquez J, Mas A, Beltrán G, Torija MJ. Efecto de los aminoácidos aromáticos y derivados sobre el crecimiento de diferentes levaduras vínicas. In: Proceedings of the Gienol Congress, Tarragona, Spain. 2015

[58] Garde-Cerdán T, Lorenzo C, Martínez-Gil AM, Lara JF, Pardo F, Salinas MR. Evolution of nitrogen compounds during grape ripening from organic and non-organic Monastrell – Nitrogen consumption and volatile formation in alcoholic fermentation. In: Nokkoul R, editor. Research in Organic Farming. London, UK: InTech Open Access; 2011. pp. 123-138

Chapter 7

Heavy Metal Residues in Milk and Milk Products and Their Detection Method

Ankur Aggarwal, Tarun Verma and Sumangal Ghosh

Abstract

Milk and milk products are an essential part of the human daily diet, and their consumption is steadily increasing. Milk is regarded as a complete food because it contains all of the macronutrients including protein, carbohydrates, fat and vitamins. Milk also has a high concentration of mineral elements (metals) such as sodium, potassium, iron, calcium, magnesium, selenium, copper and zinc. They are critical for proper body growth and maintenance but excess in these metals, particularly, heavy metals cause disturbances and pathological conditions. People nowadays are concerned about food safety issues involving microbial, chemical and physical hazards. Heavy metal residues such as cadmium (Cd), lead (Pb), arsenic (As) and mercury (Hg) pose a chemical hazard. These are the main contaminants. Heavy metals are any metallic chemical elements with a relatively high density (5 g/cc) whose levels must be monitored. Atomic absorption spectroscopy can be used to estimate the heavy metal contamination in milk and milk products.

Keywords: minerals, heavy metal, maximum residual limit, FSSAI, CODEX and Atomic absorption spectroscopy

1. Introduction

Milk is a whitish liquid containing protein, carbohydrates, fat, vitamin and trace mineral elements, which are produced by mammary gland of all mature female mammals. Milk production in India increased at a growth of 6.2% in 2020–2021 reaching 209.96 million tonnes [1]. Milk products include butter, ice cream, cheese, paneer etc. These are very important components of human diets because they contain good nutritive value and are thus widely consumed by children and adults, particularly elderly people all over the world. The advancement of industry and agriculture has resulted in the release of numerous heavy metals into the environment which is harmful to the health of both animals and humans. Animals ingest heavy metals from a variety of sources including soil, water, feed and fodder. Because the mammary gland is the most physiologically active component of an animal that resulted into heavy metals are reflected in milk (**Figure 1**). Central Pollution Control Board found that the presence of higher levels of mercury (above Environmental Protection Agency (EPA)

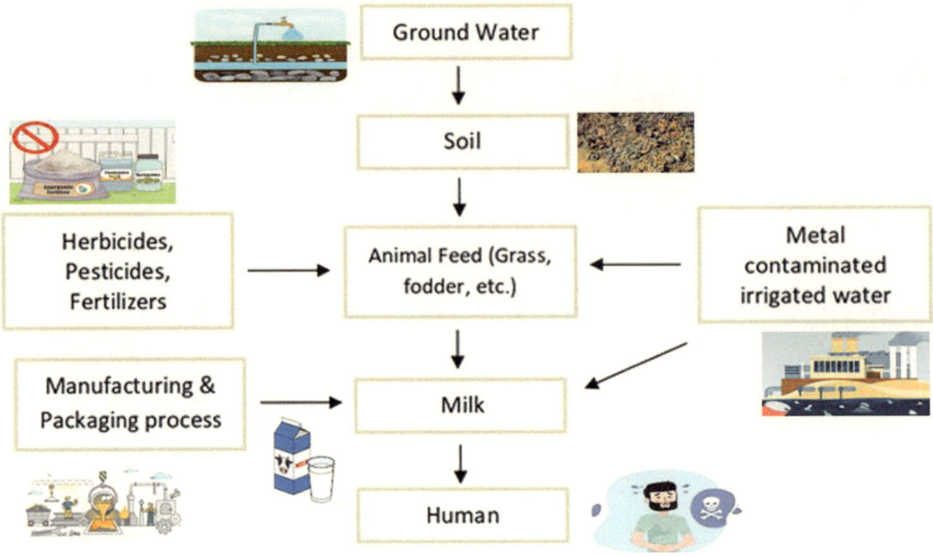

Figure 1.
Source of heavy metal in milk.

permissible limits) in water from several Indian states and its highly toxic heavy metal & poison with a long retention time in the human body poses a threat to the body. Metals of various types of minerals can be divided into two categories based on their relative abundance in our bodies. i.e., macro minerals are those that the human body requires in relatively large quantities such as sodium, potassium, chlorine, calcium, phosphorus, magnesium and sulfur whereas micro/trace minerals such as selenium, iron, zinc, copper, cobalt, fluorine, iodine, manganese and molybdenum are required in little amounts. These are found in nearly all cells of the human body where they help to maintain general homeostasis and are necessary for our bodies to function normally. The excess mineral can have a negative impact on human health e.g., daily intake of high levels of sodium can lead to hypertension (Blood pressure).

2. Heavy metals

A heavy metal is any metallic element with a relatively high density (5 g/cm^3) that is unsafe or poisonous even at low. Heavy metal is a broad term that refers to a class of metals and metalloids with atomic densities greater than 5 g/cm^3 or five times that of water [2]. Heavy metals are cumulative toxins that can cause harm even at very low levels. The toxicity of these metals is divided into two categories

a. They have really no known metabolic function but when present in the body, they disrupt the normal cellular processes which causes toxic effects in a variety of organs.

b. Bioaccumulation or the ability to accumulate in biological tissues [3].

The growth in a chemical concentration in a biological organism over time in relation to the chemical concentration in the environment is referred to as bioaccumulation. The World Health Organization (WHO), CODEX and the Food Safety and Standard

Authority of India (FSSAI) have determined metal maximum residual limit (MRL) values in food products. Heavy metals that exceed the MRL level in any food or food product harm human health.

3. Heavy metal problems in milk and milk products

Because it contains nutrients such as protein, fat, carbohydrate and minerals in which milk is considered nearly a complete food in our daily diet [4]. As industrial and agricultural processes expand, which result in the concentration of physical, chemical and biological hazards in the environment grows [5]. A significant quantity of heavy metals found in plants & animals eventually finds their way into food harming both the quality of final products and human health.

Metal levels in cow or buffalo milk are currently being examined intensively, particularly in industrialized & polluted areas of both developing and developed countries. According to reports, the basic ingredients in cow and buffalo milk are rather consistent and only alter slightly based on a variety of conditions such as lactation phase, nutrition quality and environmental conditions, all of which are primarily chemical contaminants.

Because milk and milk products are primarily consumed by infants and children, residues of lead, cadmium, arsenic and mercury are of great concern. As a result, their levels in food and food products must be monitored and controlled. Heavy metal level measurement is useful not only for determining risk to human health and assessing environmental quality but also for maintaining the high quality of final food and food products [6]. Many studies have been published on heavy metals in milk and their presence in milk and various milk products has been connected to lactating cows being exposed to pollution, consuming polluted feed and water and the manufacturing process of various milk products. Lead, mercury, arsenic and cadmium levels in milk from cows grazing in open fields in Kaduna were found to be higher than the WHO recommended limit daily intake (50 ppb) [7].

Heavy metal residues such as lead (Pb), arsenic (As), mercury (Hg) and cadmium (Cd) which pose a chemical hazard are described below.

3.1 Lead as heavy metal pollutant

Lead is one of the most dangerous metals for humans, plants and animals and it is among the most common metals in the environment due to human activities. Lead mines, coal combustion, wastewater applications, industrial waste & farmyard manure are the primary reason sources of lead in the environment [8]. Lead is a non-ferrous metal that is widely used in a variety of industries including the manufacture of plastics, storage battery alloys, ceramics, cable sheathing and even paints. The production of anti-knock compounds from petrol results in increased air pollution in the environment. Vehicle exhausts are a major source of lead contamination in the environment affecting the quality of food and food products as well as the health of animals [9]. Inhalation, ingestion and skin contact are the three main routes through which lead enters the human body system. Long-term doses of lead may cause thalassemia, pale skin, decreased muscle activity, stomach pain, vomiting, wrist joint paralysis and decreased fertility and birth abnormalities (**Table 1**). Prolonged exposure may also cause kidney damage, liver problems, nervous system damage and eventually death in humans [14]. The maximum

Heavy metals	Application	Human health consequences	Reference
Mercury	Metallurgy industries chemical manufacturing and metal finishing, use in thermometer	Memory issues increased heart rate, tremors, kidney, brain and liver damage	[10]
Arsenic	Metal plating electroplating leather, dye production	Ulcer, liver problems and kidney damage	[11]
Lead	Metal plating battery manufacturer automotive and petroleum industries	Abortion on the spur of the moment causes nervous system damage, kidney & brain damage and liver problems.	[12]
Cadim	Electroplating, fertilizers and battery manufacturing	Cancer, lung insufficiency disturbance in liver and kidney	[13]

Table 1.
Adverse effect of heavy metals on human health.

acceptable concentration of lead for milk and milk products is 250 ppb according to FSSAI and WHO.

3.2 Arsenic as heavy metal pollutant

Arsenic can be found in water that has been contaminated with industrial or agrochemical waste. Ingestion of arsenic at low doses through food or water is the main route of this metalloid into the organism with absorption taking place in the human stomach and intestines and release into the bloodstream (**Table 1**). Arsenic can be found in both natural and man-made environments. Arsenic contamination has been documented in a variety of foods and food items including tap water, air, foods and beverages (**Table 1**). Drinking water contamination is increasing as a result of industrial operations and excessive groundwater withdrawal for irrigation [15].

3.3 Mercury as heavy metal pollutant

Almost all mercury compounds are extremely toxic and can harm human and animal health even at very low levels. Mercury is subject to bioaccumulation which is the process by which organisms (including humans) absorb toxins faster than their bodies resulting in mercury levels in their bodies building up over time and causing adverse health effects in humans [16].

Human activities such as the use of fossil fuels particularly coal for heat and energy, the production of metals, cement, caustic soda, the disposal of mercury-containing waste materials etc. are the main sources of mercury in food and the food chain. Human activities that contribute to contaminated air include increased industrialization and small-scale coal burning for heat and automobiles [17]. In many state areas, mercury concentrations have increased as a result of increased atmospheric deposition which harms both humans and animals (**Table 1**).

3.4 Cadmium as heavy metal pollutants pollutant

Cadmium is a poisonous element that can harm your health. Its existence in water, soil, beverages, herbal medications, milk products and other places has gained recent notice. Phosphate fertilizers, nonferrous smelters, sewage sludge application and fossil fuel combustion are all sources of cadmium in soil and plants. According to FSSAI, the MRL level of heavy metals in milk and various milk products should not exceed 1.5 ppm.

Cadmium is used in plating, alloying, pigments, polymers and batteries among other things which is poisonous to people and animals (**Table 1**) [18].

4. Heavy metal standard for various milk products

The Food Safety and Standard Authority of India (FSSAI), CODEX and WHO tolerable weekly intake of heavy metals established standards for heavy metals in milk and milk products such as lead, arsenic, mercury and cadmium (**Table 2**).

High Sn and Ni contents of some milk products samples from this Arak, iran might be potentially hazardous to consumers [19].

A total of 65 cow and 126 buffalo milk samples were collected from various Haryana districts, covering both industrial and non-industrial areas, and it was discovered that the milk samples collected from various Haryana districts contained Pb, Cd, As, and Hg levels below the maximum contamination level, making them safe for human consumption [20].

The amount of iron, copper, manganese, zinc, lead, cadmium and chromium in cow milk yoghurt had fallen by 0.40–15% and buffalo milk yoghurt had decreased by 0.50–15% according to a study. Nickel, cobalt and tin levels in cow milk yoghurt

Heavy metals	Standard for milk and milk product		
	FSSAI (mg/l)	CODEX (mg/l)	WHO weekly tolerance toxic heavy metal intake (mg/l)
Mercury	1000	500	5
Arsenic	110	20	15
Lead	250	140	25
Cadmium	150	200	25

Table 2.
Food Safety and Standards Authority of India (FSSAI), CODEX standard and WHO Weekly tolerance intake for milk and dairy products.

Type of products	Type of heavy metal (ppb)					
	Milk	Paneer	Dahi	Cheese	Khoa	Milk powder
Lead	4.55–8.16	3.97–6.28	4.12–9.79	3.16–10.93	11.69–13.89	3.99–5.01
Cadmium	9.96–11.89	3.38–9.53	7.84–11.50	9.16–10.90	16.91–26.41	7.73–10.2
Mercury	4.48–7.23	4.23–8.53	4.04–8.04	4.87–8.68	7.46–10.68	3.34–5.55
Arsenic	4.87–8.94	3.77–8.98	4.05–11.32	5.01–7.73	14.61–21.04	6.7–9.7

Source: [21]

Table 3.
Concentration of heavy metal in milk and milk products.

were down 50–100% while buffalo milk yoghurt was down 25–50%. The level of these metals in yoghurt is dramatically reduced as a result of the high acidity and bacterial activity in the production process [4]. Another study looked at heavy metals like cadmium, mercury, lead and arsenic in milk samples collected from the Livestock Production and Management Production Department at NDRI Karnal. They discovered that higher levels of lead, cadmium and mercury in various milk products could be due to high exposure to heavy metal sources in the soil and water near hazardous waste sites while higher levels of arsenic in various milk products were mainly due to use in veterinary medicine for the eradication of tapeworms in cattle (**Table 3**) [21].

5. Reducing the contaminations

Many studies have looked at removing heavy metal pollution from many sources particularly water sources and this method can be used to disinfect a range of various dairy products mineral absorbents like smectite and Palygorsctite were used to absorb heavy metals in recent times [22]. Some other study was using sepiolite minerals and zeolites as heavy metal adsorbent materials and corrective agents [23]. Another study used a modified rice husk with different sodium bicarbonate concentrations to absorb low levels of cadmium in aquatic settings [24]. Penaud *et al.* [25] discovered that lactobacilli as probiotic agent could absorb heavy metals from products such as yoghurt.

Nurdin *et al.* [26] investigated the effect of medicinal herbs in the diet on the quantity of lead excreted in cow's milk where researcher discovering that cumin, white turmeric and mango turmeric reduced the amount of lead in various milk products by 98.36, 99.33, and 99.37% respectively.

Heavy metals bind to lactobacilli-specific proteins (LAB) and are then biologically absorbed and removed [27].

6. Analytical methods for estimation of heavy metals in various milk products

6.1 Atomic absorption spectroscopy

Atomic Absorption Spectroscopy (AAS) became available for the first time in 1962. Since then, there have been several rapid developments such as a variety of fuels and oxidants in atomic absorption and emission spectroscopy such as flame atomic absorption spectroscopy (FAAS), graphite furnace atomic absorption spectroscopy (GFAAS), vapour atomic absorption spectroscopy (CVAAS) and hydride vapor atomic absorption spectroscopy (CVAAS).

AAS which measures ppb levels in various milk product samples which are exceedingly sensitive. The spectra formed when the sample is excited with radiation from a hollow cathode lamp. Transitions to higher energy levels occur as a result of absorbing ultraviolet or visible light and then measure the amount of energy in the form of photons of light absorbed by the sample and sends the signal to the detector. In this procedure, the wavelengths of light transmitted by the sample are measured and compared to the wavelengths that passed through it originally (**Figure 2**). Atomic absorption spectroscopy is a technique for identifying an element's concentration in a sample by measuring the intensity of external absorbed radiation by a sample atom at

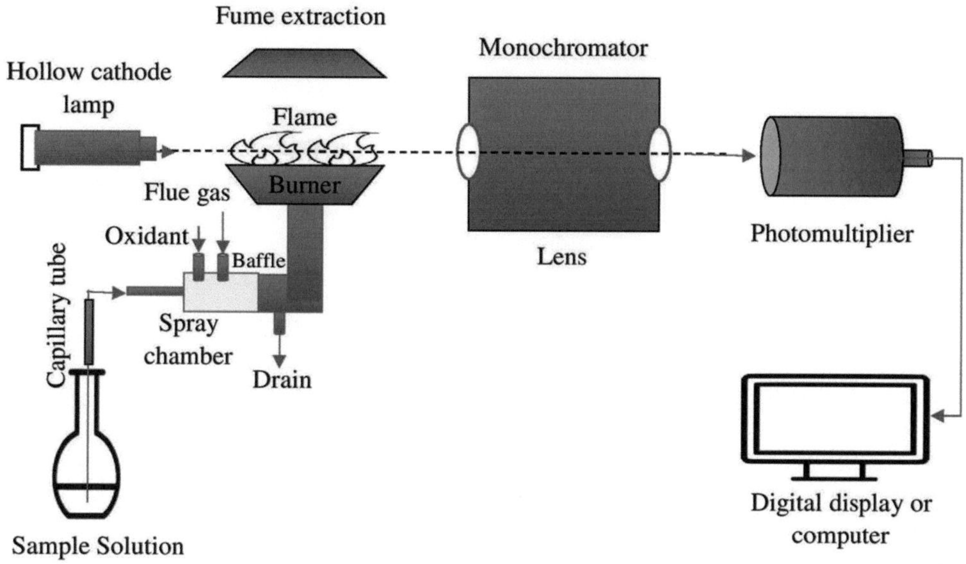

Figure 2.
Atomic absorption spectroscopy.

a wavelength characteristic of the element. The absorption of electromagnetic radiation by well-separated atoms or ions in the gaseous state is quantified using atomic absorption spectroscopy (AAS). The emission of radiation from atoms stimulated by heat or other methods is measured using atomic emission spectroscopy (AES).

6.2 Basic principle of AAS

In this system, the atoms of various elements absorb light at different wavelengths. When analyzing a sample containing a given element, light from excited atoms produces the proper wavelength combination to be absorbed by any elements present in a sample. A lamp containing atoms emits light from excited atoms resulting in a spectrum of wavelengths absorbed by any atoms in the sample is a way to find determining atom concentration in samples (**Figure 2**). Atomic Absorption Spectroscopy (AAS) atomizes the sample by converting ground state free atoms to a vapour state and passing a beam of electromagnetic radiation generated by excited atoms through the vaporized sample. The sample

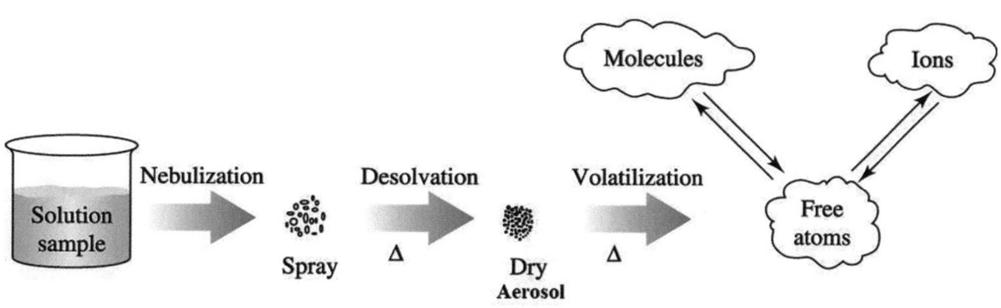

Figure 3.
Basic principle of atomic absorption spectroscopy.

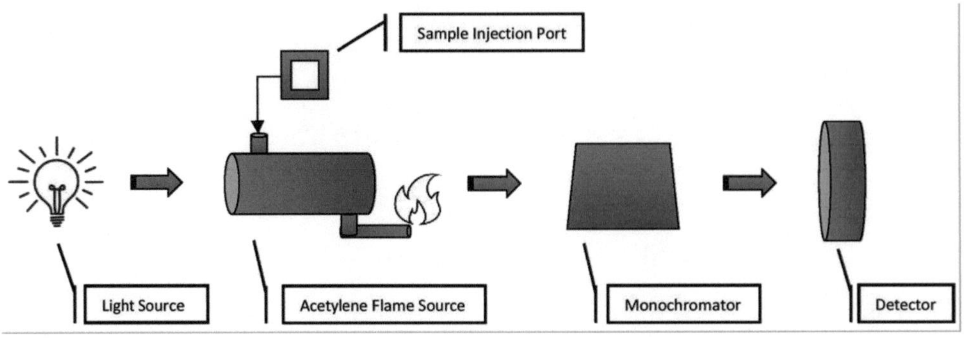

Figure 4.
Flame atomic absorption spectroscopy (FAAS).

unique atoms absorb some of the radiation because more light is absorbed the maximum number of atoms are in the vapor which transfers the signal to the detector (**Figure 3**).

6.3 Techniques of AAS

There are various techniques which can be used in AAS for the estimation of heavy metal residue in milk and milk products, which are described below:

Flame Atomic Absorption Spectroscopy is a widely used technique for estimating or detecting heavy metals, metalloids etc. in samples (ppb). A hollow cathode lamp emits radiation from a line source of the element of interest and samples are often delivered into the flame using a sprayer or nebulizer that creates minute sample droplets. The sample particles vaporize and break down into atoms, ions and electrons as the solvent in the droplets evaporates quickly. The power of the source is reduced because atoms in the sample absorb radiation released by the identical atom in the hollow cathode lamp. A monochromator is typically used to separate a spectral line of interest from any flame source background radiation. The monochromator for sodium is adjusted to pass radiation with a wavelength of 589 nm. A liquid sample is inhaled and combined as an aerosol with flammable gases in this approach (acetylene and air or acetylene and nitrous oxide). A flame with a temperature ranging from 2000°C to 3000 °C is used to ignite the mixture (depending on which fuel gas is used). An emitted light from a lamp with a cathode made of the elements to be evaluated is carried into a monochromator where it is converted to a signal and supplied to the detector via the flame (**Figure 4**).

Graphite Furnace Atomic Absorption Spectroscopy (GF AAS) – Graphite furnace atomic absorption spectroscopy is a highly sensitive spectroscopic technique for measuring various metal concentrations in aqueous and liquid samples with outstanding detection limits (ppb). GFAAS has a number of advantages over conventional analytical processes including increased sensitivity and lower limit of detection, less spectrum interference and the ability to use very small sample quantities. These all contribute to enhanced heavy metal detection or estimation accuracy. The graphite furnace is a 3000°C capable high-temperature electro-thermal atomizer system. The thermal energy of the heated graphite furnace is used to break chemical bonds inside the sample which releases free ground-state atoms capable of absorbing light energy and sending a signal to the detector & displaying the results.

Metals in solution or samples can be easily recognized using graphite furnace atomic absorption spectrophotometry (GFAAS). The approach is simple, quick and may be used to estimate metals in a variety of environmental samples including

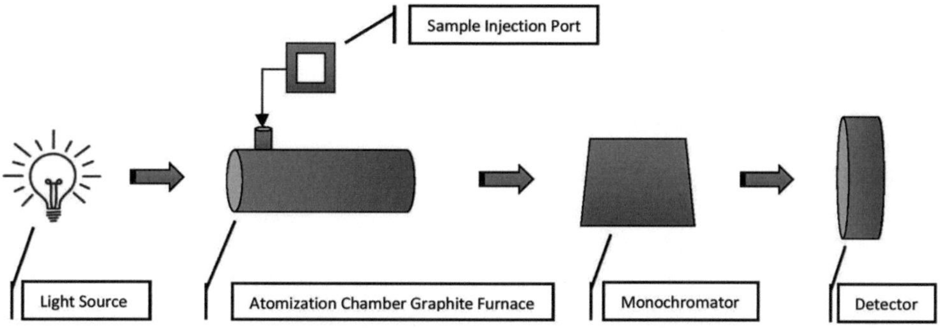

Figure 5.
Graphite furnace atomic absorption spectroscopy.

	Flame atomization	Graphite atomization
Amount of sample required	1 ml	5–20 μL
Atomization principle	Atomized through the heat of flame	Atomized by heat generated when a current is passed through a resistance bulb
Sample usage (Atomization efficiency)	Approx. 10%	Approx. 90%
Shape of adsorption signal	Stationary signal	Peak shaped signal
Reproducibility	<RSD 1%	Approx RSD 2–5%
Sensitivity	PPM Level	PPB Level
Measuring timing	Short (10–30 seconds)	Long (1–5 min)

Table 4.
Difference between flame atomic absorption spectroscopy (FAAS) and graphite furnace atomic absorption spectroscopy.

groundwater, household & industrial wastes, extracts, soils, sludge, sediments and other wastes. Except for dissolved ingredient analyses, all samples in this system must be digested before being analyzed.

Graphite Furnace Atomic Absorption Spectroscopy is similar to flame AA in that the cloud of atoms is generated by heating a tiny electrically heated graphite tube or cuvette to a temperature of 3000–3200°C. The increased atom density and tube residence length improve furnace AAS detection limits allowing detection in the ppb or level range (**Figure 5**). A comparative study between the Flame Atomic Absorption Spectroscopy (FAAS) and Graphite Furnace Atomic Absorption Spectroscopy (GFAAS) is described in **Table 4**.

Vapour Generation Atomic Absorption Spectroscopy (VG AAS) – It consists of Hydride Generation Atomic Absorption Spectroscopy (HGAAS) or Cold Vapor Atomic Absorption Spectroscopy (CVAAS). This typical approach for analyzing or estimating different mineral elements, metals and some metalloids is atomic absorption spectroscopy (AAS). However, due to interferences, poor repeatability and detection limitations, hydride generation atomic absorption spectroscopy is frequently used to analyze metalloids such as antimony, arsenic and selenium. The hollow cathode lamp, air/acetylene flame, optical system and its detector utilized in AAS are all the same in HGAAS. Materials react with sodium borohydride and hydrochloric

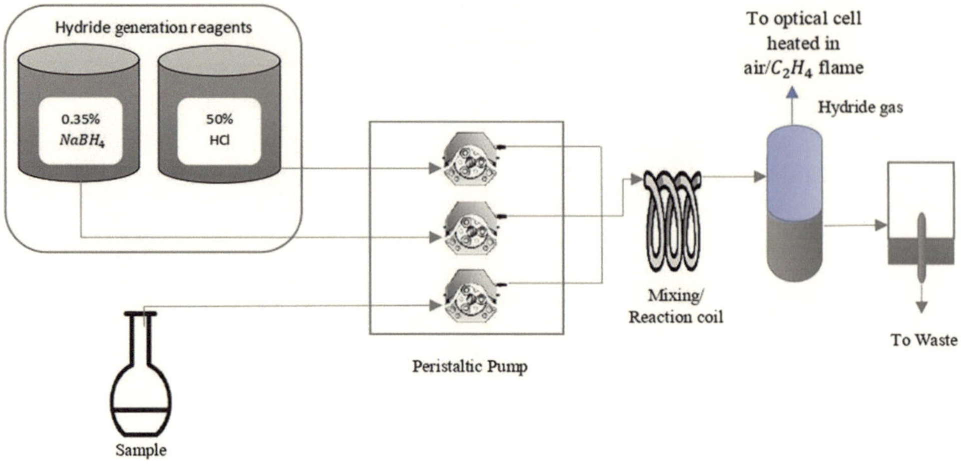

Figure 6.
Hydride generation atomic absorption spectroscopy.

acid to produce a volatile hydride in this process for example Arsenic interacts with sodium borohydride to form H_3AS (Arsenic sodium hydride). The functions of the hydride generating system are as follows:

1. Aspiration of the liquid sample followed by mixing with sodium borohydride (NaBH4) and hydrochloric acid (HCL).

2. The reaction produces a volatile hydride of the analyte metalloid.

3. Fill the optical cell system with gaseous hydride.

Atomic Absorption Spectroscopy (AAS) is a widely used technique for determining mineral elements in samples. However, some elements mostly metalloids have been developed due to interferences, low repeatability and inadequate detection methods. However, it is more expensive than atomic absorption spectroscopy (**Figure 6**). In AAS, a nebulizer is required but not in HGAAS.

7. Conclusion

Heavy metals are frequently thought to be extremely hazardous and harmful for the environment. People nowadays are concerned about food safety issues such as microbial, chemical, and physical risks. Heavy metal residues such as cadmium (Cd), lead (Pb), arsenic (As), mercury (Hg) and others are major pollutants in chemical hazards. Heavy metals do not naturally arise in milk as a consequence of human activities such as industrial and agricultural processes, but they can naturally occur in milk as a result of human activities such as industrial and agricultural processes. Polluted soils are a major source of Cd and Pb which can build up in milk through the food chain. Heavy metals have become pollutants in food for a variety of reasons resulting in a concern of health issues. Atomic Absorption Spectroscopy was used to determine the amount of heavy metal contamination in milk and milk products.

Acknowledgements

Sincere thanks are extended to the Institution of Eminence (IoE) scheme, Banaras Hindu University, Varanasi (U.P.) India for support under Incentive to Seed Grant under IoE Scheme (Devt Scheme No 6031 & PFMS Scheme No 3254).

Author details

Ankur Aggarwal, Tarun Verma* and Sumangal Ghosh
Department of Dairy Science and Food Technology, Banaras Hindu University, Varanasi, Uttar Pradesh, India

*Address all correspondence to: tarunverma@bhu.ac.in

References

[1] Economic Survey. 2022. Available from: https://www.indiabudget.gov.in/economicsurvey/ [Retrieved 31 March 2022]

[2] Sani U. Determination of some heavy metals concentration in the tissues of Tilapia and Catfishes. Biokemistri. 2011;**23**(2):73-80

[3] Al-Maylay IK, Hussein HG. Determination of some heavy metals concentrations in canned tomato paste. Research Journal in Engineering and Applied Sciences. 2014;**3**(3):216-219

[4] Enb A, Donia MA, Abd-Rabou NS, Abou-Arab AAK, El-Senaity MH. Chemical composition of raw milk and heavy metals behavior during processing of milk products. Global Veterinaria. 2009;**3**(3):268-275

[5] Farid S, Baloch MK, Ahmad SA. Water pollution: Major issue in urban areas. International Journal of Water Resources and Environmental Engineering. 2012;**4**(3):55-65

[6] Martín JR, De Arana C, Ramos-Miras JJ, Gil C, Boluda R. Impact of 70 years urban growth associated with heavy metal pollution. Environmental Pollution. 2015 1;**196**:156-163

[7] Lawal AO, Mohammed SS, Damisa D. Assessment of levels of copper, cadmium and lead in secretion of mammary gland of cows grazed on open fields. Science World Journal. 2006;**1**(1):7-10

[8] Purves D. Trace-element Contamination of the Environment. Amsterdam: Elsevier; 2012

[9] Reilly C. Metal Contamination of Food: Its Significance for Food Quality and Human Health. John Wiley & Sons; 2008

[10] Zahir F, Rizwi SJ, Haq SK, Khan RH. Low dose mercury toxicity and human health. Environmental Toxicology and Pharmacology. 2005;**20**(2):351-360

[11] Kapaj S, Peterson H, Liber K, Bhattacharya P. Human health effects from chronic arsenic poisoning–a review. Journal of Environmental Science and Health Part A. 2006;**41**(10):2399-2428

[12] Duruibe JO, Ogwuegbu MOC, Egwurugwu JN. Heavy metal pollution and human biotoxic effects. International Journal of Physical Sciences. 2007;**2**(5):112-118

[13] Bernard A. Cadmium and its adverse effects on human health. Indian Journal of Medical Research. 2008;**128**(4):557

[14] Tong S, Schirnding YEV, Prapamontol T. Environmental lead exposure: A public health problem of global dimensions. Bulletin of the World Health Organization. 2000;**78**(9):1068-1077

[15] Abernathy CO, Liu YP, Longfellow D, Aposhian HV, Beck B, Fowler B, et al. Arsenic: Health effects, mechanisms of actions, and research issues. Environmental Health Perspectives. 1999;**107**(7):593

[16] Bhan A, Sarkar NN. Mercury in the environment: Effect on health and reproduction. Reviews on Environmental Health. 2005;**20**(1):39-56

[17] Igwe JC, Nwokennaya EC, Abia AA. The role of pH in heavy metal detoxification by biosorption from aqueous solutions containing chelating

agents. African Journal of Biotechnology. 2005;**4**(10):1113-1116

[18] Jarup L. Hazards of heavy metal contamination. British Medical Bulletin. 2003;**68**(1):167-182

[19] Arianejad M, Alizadeh M, Bahrami A, Arefhoseini SR. Levels of some heavy metals in raw cow's milk from selected milk production sites in Iran: Is there any health concern? Health Promotion Perspectives. 2015;**5**(3):176

[20] Roy D, Bharathidhasan S, Mani V, Kaur H, Kewalramani N. Heavy metal contents in cow and buffalo milk samples from Haryana. Indian Journal of Animal Nutrition. 2009;**26**(1):29-33

[21] Singh M, Sharma R, Ranvir S, Gandhi K. Assessment of contamination of milk and milk products with heavy metals. Indian Journal of Dairy Science. 2020;**72**(6):608-615

[22] Farrah H, Pickering WF. The sorption of lead and cadmium species by clay minerals. Australian Journal of Chemistry. 1977;**30**(7):1417-1422

[23] Shirvani M, Kalbasi M, Shariatmadari H, Nourbakhsh F, Najafi B. Sorption–desorption of cadmium in aqueous palygorskite, sepiolite and calcite suspensions: Isotherm hysteresis. Chemosphere. 2006;**65**(11):2178-2184

[24] Shahmohammadi HZ, Moazed H, Jafarzadeh HNE, Haghighat JP. Removal of low concentrations of cadmium from water using improved rice husk. Water and Wastewater. 2008;**19**(6):27-33

[25] Penaud S, Fernandez A, Boudebbouze S, Ehrlich SD, Maguin E, Van De Guchte M. Induction of heavy-metal-transporting CPX-type ATPases during acid adaptation in Lactobacillus bulgaricus. Applied and Environmental Microbiology. 2006;**72**(12):7445-7454

[26] Nurdin E, Putra DP, Amelia T. Analysis of heavy metal lead (Pb) levels with Aas in cow's milk by giving cumin (Cuminum cyminum L.), white turmeric (Curcuma zedoaria Rosc.) and mango turmeric (Curcuma mangga Val). Pakistan Journal of Biological Sciences. 2013;**16**(21):1373

[27] Kinoshita H, Sohma Y, Ohtake F, Ishida M, Kawai Y, Kitazawa H, et al. Biosorption of heavy metals by lactic acid bacteria and identification of mercury binding protein. Research in Microbiology. 2013;**164**(7):701-709

Chapter 8

Optimization of Cassava (*Manihot esculenta* Crantz.) Fermentation Processes for Food-Secured Twenty-First Century Africa

Abosede Fawole and Adelodun Kolapo

Abstract

In mid-2021, Africa's population was more than one billion three hundred and seventy million persons and this figure is expected to reach 2.5 billion people by 2050. On the basis of regional population, Eastern, Middle and Western Africa countries are bound to continue to experience rapid rates of population growth in the coming decades. Between 1960 and 2010, the urban population of Africa increased by a factor of 8. Based on 2018 United Nation's projection, nearly 60% of African people will be living in urban areas by 2050. Of deep concerns are most Eastern, Middle, and Western African countries which will have their urban populations multiplied by a factor of 3 to 7. Similarly, the number of youth entering the labor force, as well as elderly, will continue to grow rapidly. The dramatic increase for food secured Africa occasioned by the sprawling African population, and the corresponding jobs required will be the biggest challenge most African countries will need to confront between now and 2050.The biggest challenges in this respect will be in Eastern, Middle and Western Africa where the number of youth will triple in the next 30 years. Currently, different cassava fermented products are staples in various countries of Africa. Specifically, in the past six decades, cassava has become a food security crop in sub-Saharan Africa (SSA) due to many favorable reasons. Many of these African cassava fermented products are still traditionally produced by spontaneous fermentation making the quality and safety of products uncontrollable, in addition to their incapability of mass production. Modern techniques have taken fermentation beyond preservation such that production is optimized via starter culture fermentation. These are done with a view of generating desirable organoleptic properties with quality and safety in mind. Beside this, a shift from traditional to industrial processing of cassava fermented products will generate jobs and provide enough food that will be needed to feed the projected highly urbanized future Africa. This paper advocates for the urgent need for optimization of cassava fermentation processes in Africa. However, such optimization should be characterized with multiple cultures fermentation which will generate desirable organoleptic properties, nutrition, quality and safety. Additionally, the wealth of information from genomics and proteomic era should be harnessed for improved culture performance and activities so as to improve the safety, quality and nutrient composition of cassava fermented food products indigenous to Africa. The possible impacts of such shift on food security in

IntechOpen

the twenty-first century Africa, realization of inclusive growth, poverty reduction, and achievement of economic convergence are going to be huge. It is therefore imperative for various African governments and policymakers to integrate this suggested shift into their future developmental plans so as to avoid the impeding 'demographic time-bomb'.

Keywords: Africa, cassava fermented products, food security, demography, optimization, twenty-first century

1. Introduction

In mid-2021, Africa's population was more than one billion three hundred and seventy million persons and this figure is expected to reach 2.5 billion people by 2050 [1]. The drivers for the rapid African population growth include high level of youthfulness of the population of most countries, high fertility levels, and high urban population growth [2]. On the basis of regional population, Eastern, Middle and Western Africa countries are bound to continue to experience rapid rates of population growth in the coming decades. However, the Northern and Southern Africa's share of the total African population will decrease from 21–13% and 6–3% respectively by 2050 [2].

The urban population of Africa increased by a factor of 8 between 1960 and 2010 [3]. Based on 2018 United Nation's projection, about 59% of African people will be living in urban areas by 2050 [4]. Guengant and May [2] submitted that Africa's rapid urbanization resulted in 50 cities with more than a million inhabitants and two mega cities (Cairo, 11 million and Lagos, 10.8 million) in 2010. However, these authors further estimated that in 2030 there will be four mega-cities and 11 cities of 5–10 million inhabitants on African continent while 15 mega-cities and 20 cities with a population between 5 and 10 million people are highly probable by 2050. As per the projected increase in Africa's urbanization rate, of deep concerns are most Eastern, Middle, and Western African countries which will have their urban populations multiplied by a factor of between 3 and 7.

Presently, a phenomenon known as "youth bulge" is a common experience in most African countries whereby youth aged 15–29 represent over 40% of the adult population. The attendant implication of this is that the number of youth entering the labor force will continue to grow rapidly between now and 2050. In 2010, the African and sub-Saharan Africa working age population aged 20–64 were estimated at 466 and 353 million respectively. However, it will reach 774 and 616 million respectively by 2030. By 2050, the number of 20–64 years old Africans will reach 1.097 billion. The implication of these is that between 2010 and 2050 a huge increase of labour force of 2.4–2.7 times the 2010 number will be witnessed [2]. The biggest challenge in this respect will be in Eastern, Middle and Western Africa countries where the number of youth will triple in the next 30 years. African demographic projection of the elderly up till 2050 has not been favorable as well. Almost all African countries are projected to experience dramatic increases of their elderly populations. The number of persons aged 65 years and more was estimated at 36 million in 2010. This number is expected to double and quadruple by 2030 and 2050 respectively.

Given the foregoing demographic analysis, it will be difficult to imagine the magnitude of the problems that will be associated with the sprawling African population, increasing urbanization rates, increasing "youth bulge" and the corresponding jobs required, and increasing elderly population by 2050. The dramatic increase for food secured Africa will most probably be the biggest challenge most African countries will need to confront between now and 2050.The biggest challenges in this respect will be in Eastern, Middle and Western Africa countries.

Various African countries are in different stages of demographic/fertility transition [2]. A total of 13 countries (all Southern and Northern Africa as well as island countries), accounting for 22% of the total population of the continent have completed their transition. In 41 countries, transition is still far from completion; where it ranged from "in progress transition" to "slow and irregular transition" and "very slow and/or incipient transition". Overall, 31 countries, accounting for nearly 60% of the population of the continent and 70% of the sub-Saharan Africa population can be considered as being far from completing their transition. Demographic transition is usually accompanied by epidemiological transition, i.e., a shift in health patterns from communicable to non-communicable diseases [5]. In addition, socioeconomic changes are also associated with such transition whereby the economy shifts gradually from agricultural to industrial production and eventually to a service-based economy.

The onus for a food-secured twenty-first century Africa coupled with the realization of inclusive growth, poverty reduction, and achievement of economic convergence now squarely lays on various African governments and policymakers to design future developmental plans so as to avoid the impeding 'demographic time-bomb'. In this regard, this paper advocates for the urgent need for optimization of cassava fermentation processes in Africa towards a food-secured and economically prosperous twenty-first century Africa.

2. Cassava: Africa's super crop

The Portuguese traders from Brazil introduced Cassava (*Manihot esculenta* Crantz) to sub-Saharan Africa (SSA) in the sixteenth century [6]. The crop is now produced in 40 of the 53 countries of Africa, stretching through a wide belt from Madagascar in the Southeast to Senegal and to Cape Verde in the Northwest. Current statistics indicate that African countries account for 64% of the global production of cassava and five of them are among the top 10 largest global producers. In the lead is Nigeria, who is the largest producer of cassava in the world, representing 19.4% of the global production [7].

Cassava is cultivated under a wide range of ecological and agronomic conditions of SSA. It is adaptable to relatively marginal soils and erratic rainfall conditions. It gives high productivity per unit of land and labour. The certainty of obtaining some yield even under the most adverse conditions and the possibility of maintaining continuity of supply throughout the year make the crop very adaptable to rain-fed agriculture conditions [6]. Further to these, in the future, cassava has the potential to become a promising crop that can adapt to changing climatic patterns due to its low water and soil acidity requirement compared to rice [8, 9]. The global Agro climatic suitability map for cassava under rain-fed conditions and low level of inputs is shown in **Figure 1**. Given the state of Agricultural development in Africa, majority of crop cultivation are done under rain-fed conditions and low level of inputs and it is not surprising that these scenarios translate very well to a greater portions of SSA being suitable for cassava production. It was stated earlier that the biggest challenge of impeding 'demographic time-bomb' in Africa is in the Eastern, Middle and Western Africa countries. Interestingly, these regions are also the best suited regions in Africa for cassava cultivation as shown in **Figure 1**. It then behooves the concerned African leaders to consider cassava as a key driver for food security both now and in the future Africa.

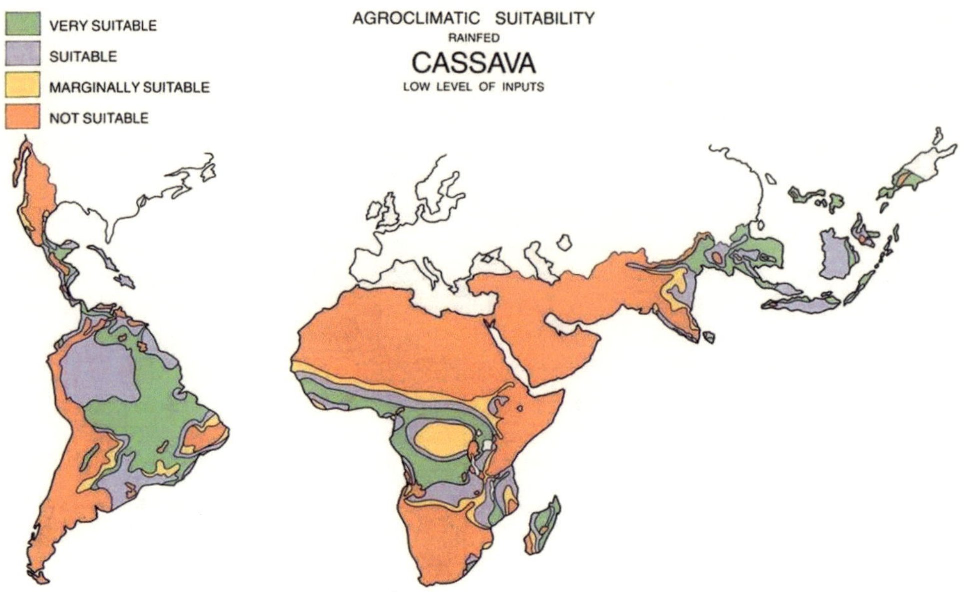

Figure 1.
Agro climatic suitability map for cassava under rain fed conditions [10].

Currently, in most African countries and cultures, cassava has become a staple crop of choice [11] as the majority of cassava tubers produced is consumed locally as traditional fermented/unfermented meals, thus making cassava an important crop by production and consumption. Cassava tuber is fermented to produce a range of different products in Africa. Among these are *garri, fufu, lafun, kokonte, agbelima, akyeke, beer, cossettes, chikwangue, kanyanga, attieke, mapanga, kivunde, Kondugbala* and *makopa*. In Asia the fermented products produced from cassava are *peuyeum, pancakes, tapai/tapey singkong, rondho royal/monyos* and *cake tape* [12–15]. Cassava is processed by a variety of methods depending on locally available processing resources, local customs and preferences. Nweke [16] reported that the five common groups of cassava products in SSA are fresh root, dried roots, pasty products, granulated products and cassava leaves.

In most SSA countries, cassava has established itself as a food security crop in the past six decades. For instance, total cassava consumption more than doubled in Africa from 24 million tonnes per year in the early 1960s to 58 million tonnes per year in the late 1990s. This large increase was due to a significant increase in per capita consumption in countries such as Ghana and Nigeria where cassava is produced as a cash crop for urban consumption. Cassava roots were the single largest source of calories in seven African countries (Angola, the Central African Republic, the Congo, the People's Republic of Congo, Ghana, Mozambique and Nigeria) having 40 percent of the population in the late 1990s, contributing an average of nearly 600 calories per person per day [16]. In a similar development, it was the second largest source of calories in another 11 countries (the Republic of Benin, Cameroon, Côte d'lvoire, Guinea, Liberia, Madagascar, Sierra Leone, Tanzania, Togo, Uganda and Zambia.) with about 25 percent of Africa's population, where it provided more than 300 calories per person per day. However, statistics from specific country indicate that many families in Congo ate cassava for breakfast, lunch and dinner thus contributing over 1000 calories per person per day or about 55 percent of the average daily calorie intake in the late 1990s [16]. It is worth mentioning that the significant contribution of cassava to SSA diets has not waned in the recent time. Detailed country data by agro-ecological zones indicates that cassava currently contributes 10–30% and 2–10% of calorie and protein supplies respectively in the diet of SSA countries in humid tropics, moist savanna, and mid-altitude regions [6, 17].

Beyond food security, cassava can potentiate a radical increase of value offering and trade status in its producing regions with a concomitant positive ripple effect on the entire economy of Africa. For instance, a study conducted in cassava growing SSA countries indicated that about 26% of cash income from all food crops in cassava-growing households was derived from sale of cassava [6]. These authors further reported that some SSA countries (Ghana and Uganda) are already taking advantage of inadequacies of Asian Exporters to satisfy the European Union market's need in relation to cassava chips. There are also some import substitution possibilities for cassava flour and industrial starch in some SSA countries.

It is on record that between year 2001 and 2007, a Presidential Cassava Initiative (PCI) project was executed in cassava-producing countries in some selected SSA countries (Nigeria, Ghana, and the Democratic Republic of Congo). The project aimed at generating about 5 billion dollars annually from exporting value-added cassava products in addition to (1) enhancing the productivity and production of cassava by increasing the area cultivated to 5 million ha, with the hope of harvesting 150 million tonnes of fresh cassava tubers annually; (2) producing 37.5 million tonnes of processed cassava products for the local and export markets [18]. A subsequent, exploratory study by [19] showed that through the PCI, the Federal Government of Nigeria created a policy measure that supported the industrialization of cassava such as 10% cassava bread, 10% bioethanol in gasoline and replacement of paraffin with ethanol gel fuel as a cooking fuel. Furthermore, there were increased investment and employment in the cassava subsector; reduction of food import bills of Nigeria; and increased cassava yield from 10.8 to 20 t/ha. Similarly, [20] submitted that PCI did not only promote cassava production, but had spillover effects on both national food production and food security.

Guengant and May [2] had submitted that the number of youth will triple in Eastern, Middle and Western Africa countries between 2010 and 2050. Beyond the quest for providing food for this teeming population, there is also a need for realization of inclusive growth, poverty reduction, and achievement of economic convergence. Given that these three regions of Africa that are of deep concern incidentally constitutes the best region in Africa which is most suitable for cassava cultivation (**Figure 1**), it is high time the policy makers in these African regions evolve cassava developmental plans that will seek to exploit the food security and socio-economic potentials that are inherent in cassava value chain.

3. Traditional processing of some cassava fermented products in Africa

Cassava (*M. esculenta*) is a plant material that is mainly processed by fermentation. It is mostly a tropical crop, but has some similarities with potato in the methods of preparation, the inedible starchy state when fresh and bland flavor when cooked. It is a food security crop in the developing world [21]. However, the main challenges faced by processors using fresh cassava root are its bulkiness and short shelf-life. It deteriorates within 3–4 days of harvest resulting in high post-harvest losses of cassava [22]. Therefore, cassava is often processed into dried forms, mostly after fermentation.

When the moisture content and water activity is lowered, it becomes durable and is shelf-stable [23]. This drought-resistant roots are fermented to produce a range of different products (**Figure 2**) which include *gari, fufu, attoupkou, lafun, kokonte, ubuswage, agbelima, akyeke, cossettes, imikembe, chikwangue, kanyanga, ikivunde, attieke, inyange, mapanga, kivunde, kondugbala, meduame-m-bong, makopa*, and *dumby* in Africa [12, 14, 24, 25].

The fermented cassava products in Africa are produced by traditional methods that exploit mixed cultures of various microorganisms and in a domestic setting with only a few small and medium scale industrial operations [25]. The techniques used are usually laborious and time consuming. Typically, such production utilize spontaneous fermentation with no implementation of good manufacturing practice (GMP) and hazard analysis and critical control point (HACCP) plans. Thus, traditional methods of processing cassava roots can result in poor and varied quality products that contain unacceptable levels of cyanide and spoilage organisms [23]. This variation in quality occurs due to initiation of fermentation by prominent microflora from the raw material, environment and water that vary according to place and time of production [12].

3.1 Gari

About 70 percent of cassava is turned into gari [24] and consumed by around 200 million people [26]. It is consumed mainly in West, Central and East Africa [24].

Figure 2.
Selected fermented cassava products: (a) gari (forfoeghgari 1.blog); (b) fufu (nigeriagalleria.com); (c) attieke (gastronomieafricaine.wordpress.com); (d) lafun (madamsabi.com); and (e) agbelima (bergamini.be).

Gari is produced by solid state fermentation to form dry, crispy, creamy-white granular product. It is made by fermenting and roasting mash produced from crushed cassava roots.

The traditional gari processing combines different stages of activity. Fresh roots are peeled, grated into pulps, and put in jute sacks for hydraulic pressing and fermentation. Some local producers use heavy stones in the place of a hydraulic jack between wooden platforms. The pulps are left in this state for 3 to 4 days, although some producers prefer a day or 2 days of solid state fermentation. It is during this stage that acidic taste is imparted to the final product [27]. Then a dewatered and fermented pulp are produced in form of lumps that are then crushed by hand. The fibers are separated from the fine pulp by sifting through a traditional sieve. The resulting small pieces are called grits which are roasted in reasonable portions using a wide frying pan with firewood as an energy source until they become dry and crispy. The processors who are mainly women are constantly in contact with smoke, heat, fume, and cyanide which affect their health over a long exposure period [28].

3.2 Lafun, cossettes, kanyanga, mapanga, and makopa

Fresh cassava roots are processed into fermented dried pulp or flour with same procedures in some African countries. The product is called lafun (Nigeria), cossettes (Democratic Republic of the Congo and Rwanda), kanyanga and mapanga (Malawi), and makopa (Tanzania). Cassava is cut into small pieces, washed, steeped in water for 3 to 4 days, drained, sun-dried and milled into powder (flour) (**Figure 3**). The characteristics of the finished products depend, to a large extent, on whether cassava roots are peeled or not before steeping in water; and whether the fermentation water is changed at intervals during fermentation [29].

Many traditional processors adopted a minimum of 72 h fermentation of cassava. Although they do not know the advantage of longer fermentation time, the 3-day fermentation period helps to reduce cyanogenic glucoside to a reasonable level. [30] noted that dietary cyanogen results in tropical ataxic neuropathy (TAN), diabetes mellitus, and can aggravate iodine deficiency disorders (IDDs). The stages of steeping and drying during traditional lafun production are typically in an unhygienic earthenware or polyethylene vessels, or rusty drums (**Figure 4**). Sun-drying the fermented cassava mash could take up to 2 weeks, depending on the weather, making the control of drying conditions very difficult [31].

There is no formal standard or quality control on this traditional processing method (**Figure 5**). That is, there is no code of hygiene, food safety or nutrition standards during production. The processors are mainly farmers. So, they harvest cassava roots, process close to the point of harvest, and sell to consumers directly or to retailers. This lack of control indicates that traditional processing of cassava has challenges associated with quality and safety. Unfortunately, it is difficult to get reliable data on the scale of what the problem of the lack of regulatory might be [23]. The fermented dried pulp of cassava is usually ground into flour to be used for the processing of secondary products, such as stiff porridge (oka), doughnuts, and cake. The stiff porridge is obtained when the fermented cassava flour is stirred in boiling water [32], while doughnuts and cake are gotten from frying dough made with dried pulp mixed with wheat flour [24].

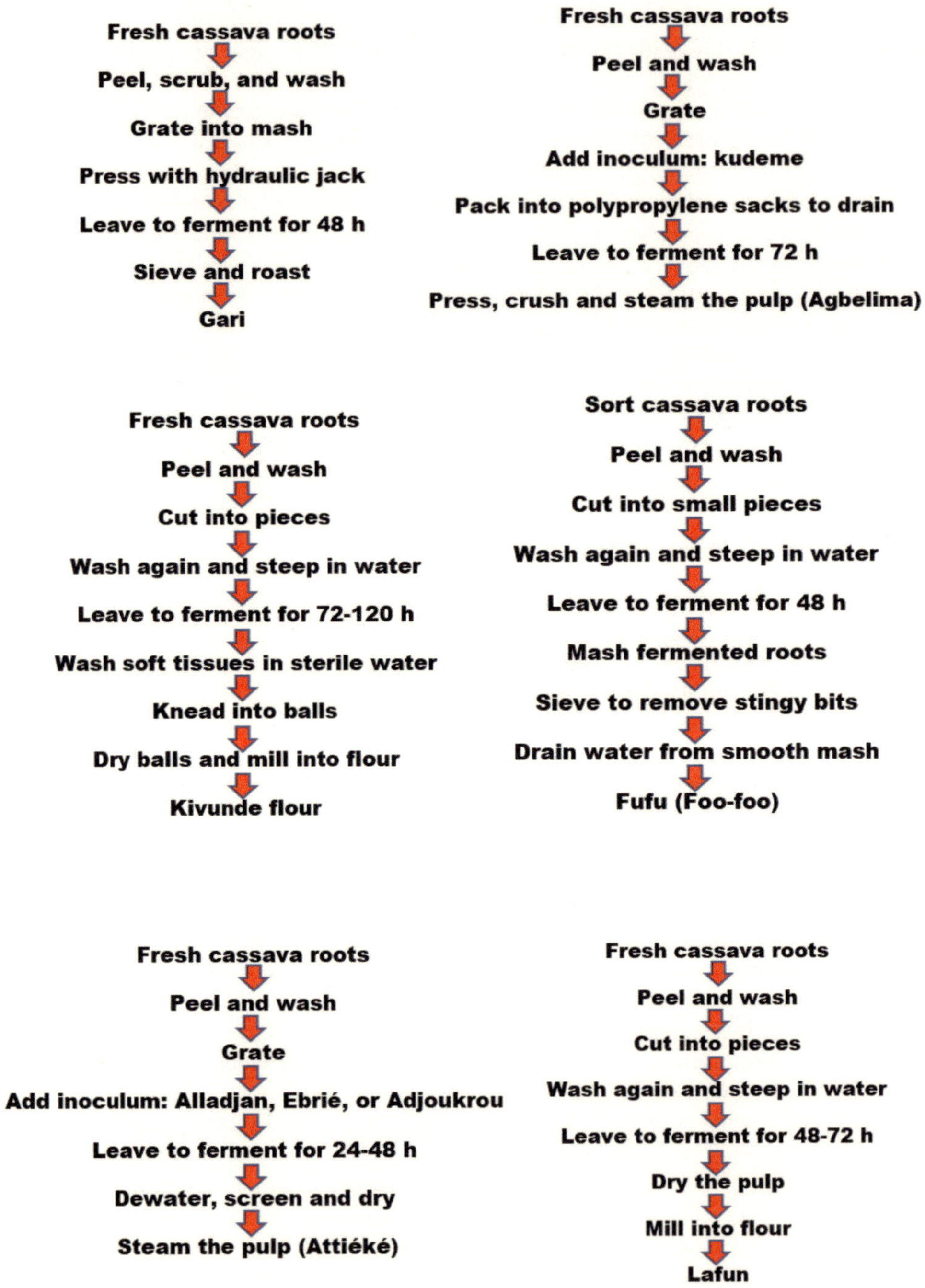

Figure 3.
Flow diagrams of the production of some fermented cassava products at ambient temperature.

3.3 Fufu

Fufu (called Foo-Foo in Central Africa) is a staple in both Central and West Africa. It is produced largely by solid-state fermentation where the cassava is not steeped in water but allowed to ferment in a solid-state under a hydraulic press [23]. Hence, the

Figure 4.
Lafun steeping and drying stages; a and c show steeping conditions; b and d are different drying sites adapted from [23].

preliminary operations are similar to that of gari. Traditional production involves peeling, washing, and cutting cassava roots into thick chunks then soaking in water contained in earthenware pots or in a slow flowing stream. The roots are left to ferment for 3 to 4 days during which they become very soft (called mash) and produce a characteristic flavor of a retted cassava product [33]. The mash is subsequently sieved after fermentation, allowed to settle within 4 h. The sediment can then be processed by cooking in boiling water [34]. Another processing method is to expel excess water from the sediment by applying heavy pressure, roll it into balls, cook it in boiling water, and pound it to a smooth paste. Fufu is a wet fermented pulp and mostly marketed in this form. Recent work on fufu is geared towards producing it in a dry form to increase shelf-stability, availability and easy transportation [35].

Figure 5.
Lafun processing environment adapted from [23].

3.4 Kivunde

Kivunde is a fermented cassava product made by heaping roots together for fermentation. It is referred to as bada in Tanzanian, and inyange in Burundi. Traditionally, kivunde is produced by spontaneous fermentation or 'back-slopping' [36]. For the back-slopping fermentation, a percentage of liquor from a prior kivunde spontaneous fermentation is added to initiate the fermentation process. The traditional fermentation of cassava into kivunde was found to be dominated by molds of the genera *Penicillium, Fusarium, Rhizopus, Cladosporium, Mucor, and Aspergillus* [12]. The traditional methods of processing cassava roots into kivunde among other products have safety issues concerning high levels of some mycotoxins [36].

3.5 Agbelima

Agbelima is a cassava fermented dough indigenous to Ghana. It is a sourdough cassava meal traditionally produced by using kudeme; an inoculum derived also from cassava roots [37]. The roots are peeled, washed, and grated with a cassava mill. The grated pulp is inoculated with kudeme, and packed into polypropylene sacks to drain and ferment for 3 days. The fermented mash is pressed, crushed and steamed [38].

3.6 Attiéké

Attiéké is the main fermented food product in Côte d'Ivoire and forms a significant part of their diet [39]. It is currently consumed in many neighboring countries like Togo, Mali and Senegal [40]. It is produced traditionally by using three different inocula: Alladjan, Ebrié, and Adjoukrou, made from boiled cassava roots [41]. Cassava roots are peeled, washed, cut in pieces, and grated. While grating, the cassava mash is mixed with any of the traditional inocula and some milliliters of palm oil. The inoculated mash is left in a covered container overnight to ferment. Excess water is squeezed out of the mash before sun-drying. The resulting granules are processed by steaming to form attiéké [24].

4. Industrialization of cassava processing in Africa: issues, challenges, and prospect

Cassava (*M. esculenta*), a shrubby tuberous plant of the Euphorbiaceae family, had been cultivated mainly for its starchy roots [42]. As an essential staple in some parts of the world, it is called different names in diverse regions. Malaysia and India named it tapioca, Francophone countries call it manioc, and in Spanish America, it is yucca. In Nigeria, it is known as ege or gbaguda by the people in the southwest; karaza or doyar kudu in the north; igari in the south, iwa unene or imidaka in the south-south; akpu, jigbo or abacha in the southeast [43].

Cassava is an important raw material for the production of many staples in Africa. Both fermented and unfermented products are derived from cassava roots across the globe. The unfermented products are high quality cassava flour (HQCF), starch, snacks, and pastries among others [23]. These unfermented products are produced from cassava type categorized as 'sweet cassava' for its low cyanide content. Thus, the processing of sweet cassava for consumption only requires a simple food preparation. The bitter variety, however, must be processed by any form of fermentation methods to attain a cyanide level that is not toxic for consumption [44].

Indigenous processors use spontaneous fermentation and back-slopping mainly as their production methods. These methods of production have classic problems associated with inconsistent product and sensory qualities. The problems are due, in part, to the microflora of the production environment, the variety of cassava used, the drying method employed, processing equipment availability and hygiene of the production plant [12]. Researchers in Food Sciences have given a great attention to upgrading traditional processes so as to optimize products and thereby increase cassava utilization.

Nigeria, Bénin, and Sierra Leone have shown notable and varying degrees of success in cassava processing at commercial level. The introduction of machines for most unit operations of cassava processing resulted in good achievement of various cassava products in the three countries. Governments policy directed towards promoting cassava subsector encouraged both small and large-scale processing industries in many Africa countries. A good example was the Nigerian Government's Cassava Initiative that started in 2001 [14]. Small- and medium-scale factories were established by some entrepreneurs to supply high-grade refined intermediate cassava products like cassava-based glucose syrup, starch and HQCF to big manufacturing industries, such as Nestlé Plc. and Cadbury. So, in Africa, the industrial utilization of cassava is not just emerging but increasing day by day. The roots have found uses in many industries processing feeds, paints, textiles, adhesives, and other chemicals [14].

Urbanization and population growth are the key factors driving the demand for cassava products in Africa. Howbeit, the commercialisation of the cassava subsector requires the inputs of responsive private sectors. The private sector inputs market in Africa is not well developed and hardly service the cassava farmers [45]. There are as much agronomic challenges as there are challenges to agro-processing. The lack of cassava stem multiplication and distribution system in Africa is a major constraint for the adoption of high yield varieties by farmers. Although medium-scale industries had been able to reduce the cassava labour bottleneck through certain mechanized technologies; cassava harvesting, peeling, and drying operations for quantity, quality, and standardized products still pose huge challenges. The chief constraint to the industrialization of cassava is perhaps the mode of drying. There is no key success to efficient dryers for mass production of cassava products at the moment. Therefore, drying process takes about 3–4 days with apparent loss in product quality [46]. Presently, high quality cassava flour processing industries in Nigeria are now leveraging on the influx of flash dryer. However, it is worth mentioning that the high initial capital expenditure for procurement, couple with the energy cost now remains a big challenge to widespread adoption of this technology.

Furthermore, there is need to separate the traditional food-oriented market from the new emerging market for industrially processed cassava. Most cassava roots grown in Africa is cultivated, processed, and traded through traditional market networks with little known about the market structure for industrial cassava [47]. Due to population growth and urbanization, there must be a paradigm shift in the traditional utilization of cassava to the opportunities in diversify cassava markets. The successful commercialization of cassava subsector also lies on the users and consumers response to the cassava transformation process for secondary product supply and new food development respectively. Other than the setback highlighted, cassava has great economical potential with its unique features as raw materials for various intermediate and final products. Improvement in cassava production and strategies for marketing products would greatly expand trade and income.

5. Food and nutrition security: Africa at a crossroad?

About two decades ago, Africa was described as a continent in crisis; as it was racked with hunger, poverty and the HIV/AIDS pandemic. It was also the region with the fastest population growth, the most fragile natural resource base and the weakest set of agricultural research and extension institutions [16]. It is disheartening that a substantial shift from this narrative is yet to be attained about 20 years after. Current report indicates that Africa is home to 43 economies with the highest poverty rates in the world and yet is the youngest continent in the world as more than 400 million of its people are aged between 15 and 35 years old [48].

Good nutrition is pivotal to the achievement of several of the Sustainable Development Goals (SDGs) such as ending poverty (SDG 1), promoting gender equality (SDG 5), ensuring quality education (SDG 4), and reducing inequalities (SDG 10) [49]. Malnutrition (in all its forms) is due to a complex set of interacting factors, including the inadequate, unbalanced or excessive consumption of the macronutrients that provide dietary energy (carbohydrates, protein and fats) and micronutrients (vitamins and minerals) which are essential for physical and cognitive growth and development [50]. Multiple burden of malnutrition may exist in the form of undernutrition and micronutrient deficiencies, but in addition, overweight and obesity are emerging as significant health concerns in a number of countries.

Globally, the prevalence of undernutrition (PoU) has remained at 10.8 percent over 2017 and 2018. In 2018, there are 822 million undernourished people in the world, up from 812 million in 2017 and 797 million in 2016.Though it is on record that the prevalence of undernourishment in Africa fell from 24.5 percent in 2000 to 18.2 percent in 2014, but then started rising to 20 percent of the continent's population, or 256 million people in 2018, with SSA having a share of 239 million undernourished people [50]. However, recent data indicates that Southern Asia and Sub-Saharan Africa respectively had PoU of 15.8 percent (257.3 million people) and 24.1 percent (234.7 million people) in 2019 [51]. In another development, out of the twenty countries affected with multiple micronutrient deficiencies globally, eighteen were from sub-Saharan Africa and two from Asia [52].

Undernutrition comes with considerable economic costs including lower cognitive skills and school attainment, and impaired physical development, which can reduce productivity in adulthood. Estimates from selected African countries (Egypt, Ethiopia, Swaziland and Uganda) indicate that productivity losses ascribed to adults, who suffered from stunting in their childhood, are huge. For example, annual losses of 3.7, 4.7, 0.092 and 0.899 billion US dollars, equivalent of 1.9, 16.5, 3.1 and 5.6% of GDP, have been estimated for Egypt, Ethiopia, Swaziland and Uganda respectively [53]. In recognition of the overwhelming economic burden of malnutrition, the WHO Member States in 2012 adopted a set of global nutrition targets. The six interlinked WHA global nutrition targets for 2025 are: achieving a 40 percent reduction in the number of children under 5 years who are stunted, achieving a 50 percent reduction of anemia in women of reproductive age, achieving a 30 percent reduction in low birth weight, ensuring that there is no increase in childhood overweight, increasing the rate of exclusive breastfeeding in the first 6 months up to at least 50 percent, and, reducing and maintaining childhood wasting to less than 5 percent. However, overall progress towards these WHA global nutrition targets remains unacceptably slow in Africa, as out of 54 African countries: 7 are on course to meet the target for stunting: 0 are on course to meet the target for anemia in women of reproductive age, 20 are on course to meet the target on overweight, 1 are on course to meet the target on exclusive breastfeeding, while 13 are on course to meet the target on wasting [50]. With these statistics, Africa is quite at a crossroad of food and nutrition insecurity.

In Africa, three major drivers of hunger and food insecurity are climate change, conflict and economic slowdowns and downturns [50]. In consideration of the fact that evolving coping strategies against the imminent consequences of climate change is more needed in Africa both now and in the nearest future, increased utilization of climate-smart crop is perhaps one of the needed approach to fight food and nutrition insecurity. In Africa, roots and tubers are one of the most consumed food groups and cassava as a climate-smart tuberous crop is an ideal crop to fight the scourge of malnutrition given its high rate of consumption by rural and urban Sub-Sahara African populations. Although cassava does not have a high nutritive value, especially for certain micronutrients and protein, the crop has benefitted from several initiatives focused on increasing its micronutrient load through biofortification. For instance, organizations like HarvestPlus have made great strides towards improving the nutritive value of cassava through the development of vitamin A-rich cassava and its promotion for adoption into the food systems of several African countries like Nigeria, Ghana, Cameroon, and DRC. In addition, food-to-food fortification using soybean flour/residue has the potential to produce cassava-based foods that could be used to fight macronutrient and micronutrient deficiencies in African countries where cassava products are staples [54, 55]. The onus therefore lays on African policy makers, especially in the Sub-Sahara Africa, to leverage on the potentials of value-added cassava-based products in addressing the twenty-first century food security concerns in Africa.

6. Optimization of cassava fermentation process: An imperative for twenty-first century food-secured Africa

As stated earlier, the dramatic increasing need for food secured Africa occasioned by the sprawling African population, and the corresponding jobs required, will be the biggest challenges most African countries will need to confront between now and 2050. In addition, the projected increase in urbanization rates and increased aged/elderly population are "demographic time bombs" that will further worsen food and nutrition security situation in most African countries in the coming decades if proactive steps are not taken to shift from traditional cottage level to mass scale food processing. While evolving future developmental plans so as to avoid this impeding 'demographic time-bomb', it is imperative for various African governments and policymakers to learn from European history, especially in relation to industrialization of food processing.

Food processing using fire and cooking has occurred since pre-historic times. However, more complex forms of food processing such as baking of bread, production of cheese, wine production, sun-dried or vinegar-pickled vegetables production, and salted or smoked meat production emerged in ancient and medieval times [56]. Historically, various factors are responsible for processing foods. These include impossibility to eat fresh food, seasonality of crop production, crop failures, wars and long sea voyages. However, the aims of modern processing are manifold, and include the prolongation of shelf-life, ensuring safety, improving palatability, increasing variety, improving nutritional value and increasing convenience [57]. Mass scale food processing (producing foods in large amounts) was introduced during the industrial revolution in the eighteenth and nineteenth century, starting with the advent of canned and pasteurized foods [56]. In the first half of the twentieth century, Europe was ravaged by malnutrition (undernutrition), caused by poverty, an economic depression and two catastrophic world wars [57, 58]. As a result, mass food production focused on sustaining the European population; reducing foodborne diseases, malnutrition and nutrient deficiencies by providing protein-rich, energy-dense and fortified foods (with vitamins) that were accessible to all [56].

It is important to envision the impact of the demographic projections for Africa up till 2050 on African food and nutrition security situation in the middle of this twenty-first century. This will be helpful to appraise the needed imperatives that must be factored in while evolving a sustainable food and nutrition security road map for Africa for the rest part of this century. Rapid population growth will demand more food to be produced using the same size of arable land. Desert encroachment and climate change and their accompanied consequences will constitute major problems for food production, except innovative ways of sustainable agricultural practices are adopted. Increase urbanization rates will limit availability and accessibility of traditionally produced food for the great majority that live in urban environment. Increased poverty rate, economic slowdown or downturn will definitely limit food accessibility of the poor majority. Most often than not, children, women and elderly are the most vulnerable groups whenever there is a food crisis. This scenario will definitely be sacrosanct amid the looming possible food crisis in Africa build up to 2050.

As stated earlier, the biggest challenges of impeding 'demographic time-bomb' in Africa is in the Eastern, Middle and Western Africa countries. Interestingly, these regions are also the best suited regions for cassava cultivation in Africa as shown in **Figure 1**. Kolapo and Sanni [59] submitted that utilization of locally sourced food material must be of primary consideration in the quest of any nutritional development programme. It then behooves the concerned African leaders to consider cassava as a key driver for food security both now and in the future Africa. This will be consistent with the earlier description of cassava as Africa's super crop.

The cassava fermentation process varies from one region to another, both in Africa and other part of the globe. However, these fermentation techniques in cassava processing are broadly categorized into solid-state and submerged fermentation. In solid-state fermentation, cassava root is not soaked in water whereas in submerged fermentation processes, cassava roots are soaked in water for the duration of fermentation. Gari, Attieke, and Injera are produced by the former technique [60–62] while Fufu, Pupuru, and Chikwangue are produced using the later technique [63–65]. Microorganisms of various groups such as lactic acid bacteria (LAB), yeasts, molds and *Bacillus* strains, among many others, are involved in the fermentation of cassava-based foods indigenous to different parts of the world. The sources of these microorganisms are usually raw ingredients, the traditional utensils used for the processes, water used for the processing and the immediate atmosphere. Some genera/species of microorganisms that have been reported concerning various cassava fermented foods in Africa are stated in **Table 1**. During fermentation of cassava products, these microorganisms have been linked to the following roles: cyanide and antinutrients reduction, protein enrichment, food preservation, texture improvement, aroma and flavor change [66–68].

The fermented cassava products in Africa are produced by traditional methods that exploit mixed cultures of various microorganisms and in a domestic setting with only a few small and medium scale industrial operations [25]. These traditional methods of processing cassava roots do result in poor and varied quality products that contain unacceptable levels of cyanide and spoilage organisms [23]. Though, the earliest production of cassava fermented foods was based on spontaneous fermentation due to the development of the microflora naturally present in the raw material, there are current attempts to optimize cassava fermentation processes using starter cultures. Padonou and co-workers [32] assessed the role of *Saccharomyces cerevisiae* 2Y48P22, *Lactobacillus fermentum* 2L48P21, *Lactobacillus plantarum* 1L48P35 and *B. cereus* 2B24P31 in root softening and the overall organoleptic quality of Lafun in a quest to develop a suitable starter culture for the standardized production of Lafun. Recently, [23] evaluated the impact of the developed starters (*Weissella koreensis* (2 strains), *Lactococcus lactis* and *Leuconostoc mesenteroides*), singly and in combination, on Lafun nutritional quality, rheological properties, volatile flavor profile, NMR metabolite profile, cyanide reduction and microbial load using three cassava varieties (Bitter: IBA30527; Fortified: IBA011371; and Sweet: TMEB117). In a related development, [69] investigated a possible nutritional enrichment of *Lafun* using *Lactobacillus plantarum* and *Saccharomyces boulardii* as starter microorganisms.

While developing starter cultures for gari fermentation, [26] recommended the inclusion of *L. plantarum, L. fallax* and *Lactobacillus fermentum species* in a mixed culture as starters for gari production because of their huge linamarase activity, fast acid production and production of antagonistic substances like bacteriocins and hydrogen peroxide. In addition, [70] submitted that these three species had a beneficial property of high sugar fermentation profile for indigestible stachyose and raffinose sugars. In another development, [71] evaluated the effect of *L. plantarum* strains in fufu production. The outcome of their study indicated that fufu, produced using *L. plantarum* as the starter culture, had more desirable pasting quality and flavor in comparison to the traditionally produced samples. Molds of the genera *Penicillium, Cladosporium, Rhizopus, Mucor, Aspergillus and Fusarium* were found to dominate the traditional fermentation of cassava into kivunde [12]. However, [36] reported that this traditional method has safety issues concerning high levels of some mycotoxins. Therefore, the potential of *L. plantarum* to improve the quality and safety of this product was demonstrated by [12]. Bouatenin and co-workers [72] used LAB strains (*L. plantarum* and *L. mesenteroides*) among other microbes as starters, both singly and in combination for attiéké production. They reported that acid production responsible for sourness and detoxification of the products was the job of LAB strains. Similarly, Agbelima, like attiéké, is a sourdough cassava meal in which souring was achieved by LAB such as *L. mesenteroides, Lactobacillus brevis* and *L. plantarum* [73].

Though these optimization attempts resulted in positive flavor, nutritional and physical properties and cyanide reduction, thereby advancing the development of these starter cultures and providing a basis for further research required for their development and adoption; there is still a long way to go in the industrialization of the production process of these Africa cassava fermented foods. It is incontrovertible that the application of molecular technologies with emphasis on the application of novel sequencing technologies to generate phylobiomes, metagenomes and metatranscriptomes for microbial community profiling that complement culturing studies has greatly facilitated monitoring of fermentation ecosystems and characterization of the microbial species in the past two decades. Such technologies involved the use of High-Throughput-Sequencing (HTS) techniques such as whole-genome sequencing, amplicon bases meta-taxonomic approaches (such as 16S rRNA sequencing), shotgun metagenomics, and (meta) transcriptomics, and rely on downstream

Food product	LAB and other bacteria	Yeasts	Molds	Countries of production
Gari	*Bacillus subtilis; Bacillus coagulant; Bacillus species; Lactobacillus plantarum; Lactobacillus fermentum; Lactobacillus brevis; Lactobacillus pentosus; Lactobacillus acidophilus; Lactobacillus sp; Leuconostoc Alcaligenes; Leuconostoc fallax; Corynebacterium manihot; Corune bacterium species, Pseudomonas mesenteroides, Weissellapara mesenteroides; Corynebacterium; Bacteroides sp; Actinomyces sp*	*Saccharomyces fragilis; Saccharomyces cerevisiae; Saccharomyces rouxii; Geotrichum candidum*		West, Central, and East Africa countries
Ikivunde	*L. plantarum; L. brevis; L. fermentum; Leuconostoc mesenteroides*	*Geotrichum candidum*		Burundi, Rwanda
Chikwangue	*Lactococcus lactis; Leuconostoc sp; Lactococcus plantarum; L. plantarum; Clostridium spp.*			Main central Africa countries
Inyanga			*Aspergillus oryzae, A. fumigatus, Penicillium chrysogenum, Rhizopus stolonifera, Mucor spp*	Burundi
Attieke	*Lactobacillus plantarum; Lactobacillus; fermentum; Lactobacillus cellobiose; Lactobacillus brevis; Leuconostoc mesenteroides; Lactobacillus sp; Bacillus sp, Bacillus sphaericus; Bacillus brevis, Bacillus; coagulants, Enterococcus faecium*	*Candida krusei; Kloeckera japonica; Saccharomyces cerevisiae*		Bénin, Mali, Sénégal, Togo

Food product	LAB and other bacteria	Yeasts	Molds	Countries of production
Fufu	*Lactobacillus callobiosus L. bulgaricus, L. brevis, L. plantarum Leuconostoc mesenteroides*	*Candida famata*		Nigeria
Pupuru	*Lactobacillus plantarum Lactobacillus fermentum Cryptococcus humicola Leuconostoc spp. Corynebacterium pyrogenes*	*Geotrichum capitatum Candida famata Saccharomyces cerevisiae Geotricum candidum*		South western Nigeria and west Africa

Table 1.
The main microorganisms associated with fermentation in some cassava-based food (Adapted from [66]).

bioinformatics analysis [74]. In this regard, an increasing number of fermented foods of Asian origin were analyzed with these techniques [75]. However, in Africa, whole-genome sequencing was initially used for the analysis and surveillance of foodborne pathogens [76].

While fermented food products around the world are increasingly being studied using shotgun metagenomic techniques, very few studies have explored the use of shotgun sequencing in African fermented foods, with some exceptions being "Nunu" (Ghana), "Kokonte" (Ghana/Togo) and "Wagashi" (Benin) [77]. Given that, shotgun metagenomic sequencing facilitates deeper insights into the microbiome, allowing strain-level identification, functional annotation including carbohydrate pathways and bioactive molecule production (such as bacteriocins), and the assembly of high quality genomes in the form of metagenome-assembled genomes (MAGs) [78, 79], there is an urgent need to harness the traditional culture-based and HTS techniques to characterize and optimize the microbiome of African cassava fermented foods, from commercial to nutritional and health-promoting aspects. This will require a huge investment on the part of the leaders of concerned African countries.

Most food fermentation processes depend on mixtures of microbes which act in concert to produce the desired product characteristics. Brenner and co-workers [80] posited that this can be explained by two features. First, members of the consortium communicate with one another by trading metabolites or by exchanging molecular signals. Second, division of labour exists between the members of the consortium leading to an overall output that can only be explained by combining tasks performed by constituent individuals or sub-populations. The best well-known example is the proto-cooperation between *Streptococcus salivarius* subsp. *thermophiles* and *Lactobacillus delbrueckii* subsp. *bulgaricus* in yogurt fermentation with clear links to product functionality [81]. Another example of proto-cooperation is the development of a highly efficient fermentation process using a co-culture of *L. plantarum* SM39, and *Propionibacterium freudenreichii* DF13 for folate and vitamin B12 production [82, 83]. In a recent alkaline fermentation of indigenous African fermented foods, [84, 85] documented the use of *Bacillus subtilis* LB3, *Staphylococcus xylosus* SAU3 and *L. mesenteroides* ssp. *cremoris* LAB5 for controlled fermentation of soybean-daddawa, in which the naturally fermented sensory attributes of soybean daddawa was nearly replicated in a controlled setting, thus holding a great promise for subsequent industrialization of soybean daddawa production. Given the foregoing scenarios, optimization of cassava fermented foods indigenous to Africa should be characterized with co-cultures fermentation which will generate desirable organoleptic, nutritional, quality, and safety properties.

While contemplating industrialization of cassava fermentation process in Africa, two possibilities that could be leveraged upon are backslopping (in the short term) and the use of functional starters (in the long term). Backslopping involves inoculation of the raw material with a small quantity of a previously performed successful fermentation, which eventually results in dominance of the best adapted strains. Functional starter cultures are starters that possess at least one inherent functional property which can contribute to food safety and/or offer one or more organoleptic, technological, nutritional, or health advantages [75]. Backslopping is still in use, for instance in the production of sauerkraut and sourdough, and particularly for products for which the microbial ecology and the precise role of successions in microbial population are not well known [86]. It also represents a cheap and reliable preservation method in less developed countries [87], whereas in Western countries the use of starter culture in large-scale production of fermented foods has become an important routine procedure of the food industry [88].

African leaders in 2013, through the 50th Anniversary Solemn Declaration during the commemoration of the Fiftieth Anniversary of the Organization of African Unity (OAU), while acknowledging past successes and challenges, rededicated themselves to the continent's accelerated development and technological progress. They emphasized a guiding vision "to build an integrated, prosperous and peaceful Africa, driven and managed by its own citizens and representing a dynamic force in the international arena", and forthwith identified seven aspirations to serve as pillars for the continent in the foreseeable future. Agenda 2063 was subsequently designed to translate these aspirations into concrete objectives, milestones, goals, targets and actions/measures. The first aspiration of the Agenda seeks to obtain prosperous Africa based on inclusive growth and sustainable development. The policy document defined a prosperous Africa as a continent with a high standard of living, quality of life and well-being; whose citizens are healthy and well-nourished, subsequent to a robust food and nutrition strategy initiatives. A prosperous Africa is also expected to be well educated and be a key player of knowledge economy and become a Science, Technology and Innovation hub. As a prosperous Africa, the continent is expected to adopt modern Agriculture for increased productivity and production. In order for Africa to attain prosperity status, it was envisioned that Africa must develop coping strategies against the imminent challenges of climate change. The sixth aspiration of Agenda 2063 envisions an Africa whose development is people-driven, relying on the potential of African people, especially its women, youth, and children. These are to be achieved by attainment of full gender equality in all spheres of life, as well as rigorous youth and children empowerment.

It is on record that African countries account for 64% of the global production of cassava and five of them are among the top 10 largest global producers [7]. Africa must therefore know what must be done to this 'cassava resource' to achieve economic prosperity as well as food and nutrition security, since utilization of locally sourced food material must be of primary consideration in the quest of any nutritional development programme [59]. The foregoing analysis encapsulated in this paper has clearly shown what must be done in this regard.

7. Conclusion

The biggest challenge of impeding 'demographic time-bomb' in Africa is in the Eastern, Middle and Western Africa countries. Interestingly, these regions are also the best suited regions for cassava cultivation in Africa It then behooves the concerned African leaders to consider cassava as a key driver for food security both now and in the future Africa. This is consistent with the earlier description of cassava as Africa's super crop.

Demographic transition is usually accompanied by both epidemiological and socioeconomic changes whereby the economy shifts gradually from agricultural to industrial production and eventually to a service-based economy. Incidentally, 31 African countries, accounting for nearly 60% of the population of the continent and 70% of the sub-Saharan Africa population can be considered as being far from completing their transitions. The projected increase in population figures, urbanization rates and increased aged/elderly population are "demographic time bombs" that will further worsen food and nutrition security situation in most African countries in the coming decades if proactive steps are not taken to shift from traditional cottage level to mass scale food production and processing.

While evolving future developmental plans so as to avoid this impeding 'demographic time-bomb', it is imperative for various African governments and policymakers to learn from European history, especially in relation to industrialization of food processing. In the twentieth century, political, social, and economic changes, and scientific and technological advances moved at an ever increasing pace and impacted on the European food processing industry and influenced the ways foods were processed and marketed. As a result, mass food production focused on sustaining the European population; reducing foodborne diseases, malnutrition and nutrient deficiencies by providing protein-rich, energy-dense and fortified foods (with vitamins) that were accessible to all.

If Africa must contain the impending demography-associated food crisis in the coming decades, increased food production and industrialization of the processing of produced food crops is never an option but a necessity. As Africa's super crop, cassava has a great role to play in this respect, as fermented cassava products have become prominent staples across many SSA countries. There is therefore an urgent need to harness the traditional culture-based and HTS techniques to characterize and optimize the microbiome of African cassava fermented foods, from commercial to nutritional and health-promoting aspects. This will require a huge investment on the part of the leaders of concerned African countries. However, given the immense benefits of such project and the expectations of Agenda 2063, African leaders must begin to put their resources where their mouths are so as to ensure that the twenty-first century Africa is food-secured.

Acknowledgements

This section of your manuscript may also include funding information.

Conflict of interest

The authors declare no conflict of interest.

Author details

Abosede Fawole[1] and Adelodun Kolapo[2]*

1 Biology Department, The Polytechnic, Ibadan, Nigeria

2 Biological Sciences Department, Augustine University, Ilara-Epe, Lagos State, Nigeria

*Address all correspondence to: adelodun.kolapo@augustineuniversity.edu.ng

References

[1] Statista, Total population of Africa 2000-2021. 2022. Available from: https://www.statista.com/statistics/1224168/total-population-of-africa/ [Accessed: 2022-03-02]

[2] Guengant JP, May JF. Mint: African demography. Global Journal of Emerging Market Economies. 2013;**5**:215-267. DOI: 10.1177/0974910113505792

[3] United Nations. World Urbanization Prospects: The 2011 Revision. New York, NY: United Nations, Department of Economic and Social Affairs, Population Division; 2012. Available from: https://www.un.org/en/development/desa/population/publications/pdf/urbanization/WUP2011_Report.pdf Accessed: 2022-03-02

[4] United Nations. World Urbanization Prospects: The 2018 Revision. New York, NY: United Nations, Department of Economic and Social Affairs, Population Division; 2019. Available from: https://population.un.org/wup/Publications/Files/WUP2018-Report.pdf Accessed: 2022-03-02

[5] May JF. World Population Policies: Their Origin, Evolution, and Impact. New York NY: Springer; 2012

[6] Spencer SC, Ezedinma C. Cassava cultivation in sub-Saharan Africa. In: Hersey CH, editor. Achieving Sustainable Cultivation of Cassava. 1st ed. Vol. 1. London: Burleigh Dodds Science Publishing; 2017. pp. 1-26. DOI: 10.4324/9781351114264

[7] Adeoye A. Mint: Cassava: A key driver for food security in Africa. Sahel Quarterly. 2021;**28**:4-6

[8] FAO. Save and Grow: Cassava A Guide to Sustainable Production Intensification. Rome: FAO; 2013

[9] Khumaida N, Ardie SW, Sopandie D. Mint: Influence of agro-ecology on growth and performance of several potential mutants of Cassava. Procedia Environmental Sciences. 2016;**33**:70-77

[10] FAO Agroclimatic Suitability Rainfed Cassava Low Level of Inputs. 2022. Available from: https://www.fao.org/3/t8300e/t8300e0q.jpg [Accessed: 2022-03-02]

[11] CGIAR. What is the True Impact of Improved Cassava Varieties in Nigeria? Brief No. 64. 2018. Available from: https://cas.cgiar.org/sites/default/files/pdf/ispc_brief_64_impact_cassava_nigeria.pdf> [Accessed: 12 November 2021]

[12] Kimaryo VM, Massawe GA, Olasupo NA, Holzapfel WH. The use of a starter culture in the fermentation of cassava for the production of "kivunde", a traditional Tanzanian food product. International Journal Food Microbiology. 2000;**56**(2-3):179-190

[13] Echebiri RN, MEI E. Mint: Production and utilization of cassava in Nigeria: Prospects for food security and infant nutrition. PAT. 2008;**4**(1):38-52

[14] Sanni LO, Onadipe OO, Ilona P, Mussagy MD, Abass A. Successes and Challenges of Cassava Enterprises in West Africa : A Case Study of Nigeria. Bénin , and Sierra Leone, Ibadan: IITA; 2009. p. 19

[15] Akely PMT, Azouma OY, Amani NG. Mechanical pressing and semolina preparation from fermented cassava

paste during "attiéké" (yucca flour) processing. Journal of Food Engineering. 2010;**101**(4):343-348. DOI: 10.1016/j.jfoodeng.2010.07.011

[16] Nweke FI. The Cassava transformation in Africa. In: A Review of Cassava in Africa with Country Case Studies on Nigeria, Ghana, The United Republic Of Tanzania, Uganda and Benin. Proceedings of the Validation Forum on The Global Cassava Development Strategy. IFAD, FAO; 2005

[17] Food and Agriculture Organization of the United Nations (FAO). FAOSTAT Food Balance Sheets. Rome: Food and Agriculture Organization of the United Nations (FAO). 2014

[18] Sanogo D, Adetunji O. Presidential initiatives on cassava in Africa: Case studies of Ghana and Nigeria. NEPAD Pan African Cassava Initiative (NPACI). Partnerships in Support of CAADP, IITA. Ibadan: NEPAD; 2008

[19] Ohimain EI. A decade (2002-2012) of presidential intervention on Cassava in Nigeria; the successes and challenges. Asian Journal of Agricultural Extension, Economics and Sociology. 2015;**6**:185-193

[20] Donkor E, Onakuse S, Bogue J, ILR C. The impact of the presidential cassava initiative on cassava productivity in Nigeria: Implication for sustainable food supply and food security. Cogent Food and Agriculture. 2017;**3**:1. DOI: 10.1080/23311932.2017.1368857

[21] Sriwana IK, Erni N. Supply chain analysis of cassava agroindustry to improve national food security. In: Industrial Engineering and Management. 2015. pp. 1-7 Available from: https://foodiepelago.files.wordpress.com/2013/10/gernas-singkong-sejahtera-b.jpeg

[22] Saravanan R, Ravi V, Stephen R, Thajudhin S, George J. Post-harvest physiological deterioration of cassava (*Manihot esculenta*)—A review. Indian Journal Agricultural Science. 2016;**86**(11):1383-1390

[23] Fawole AO. Selection of Lactic Acid Bacteria for Use as Starter Cultures in Lafun Production and Their Impact on Product Quality and Safety. thesis ed. Reading: University of Reading; 2019

[24] Flibert G, Abel T, Aly S. African cassava traditional fermented food: The Microorganism's contribution to their nutritional and safety values–A review. International Journal of Current Microbiology and Applied Science. 2016;**5**(10):664-687

[25] Huch (née Kostinek) M, Hanak A, Specht I, Dortu CM, Thonart P, Mbugua S, et al. Use of Lactobacillus strains to start cassava fermentations for gari production. International Journal of Food Microbiology. 2008;**128**(2):258-267. DOI: 10.1016/j.ijfoodmicro.2008.08.017

[26] Kostinek M, Specht I, Edward VA, Schillinger U, Hertel C, Holzapfel WH, et al. Diversity and technological properties of predominant lactic acid bacteria from fermented cassava used for the preparation of Gari, a traditional African food. Systematic and Applied Microbiology. 2005;**28**(6):527-540

[27] Hahn SK. An overview of African traditional cassava processing and utilization. Outlook on Agriculture. 1989;**18**:110-118

[28] James B, Okechukwu R, Abass A, Fannah S, Maziya-Dixon B, Sanni L, et al. Producing Gari from Cassava: An illustrated guide for smallholder cassava processors Ibadan: International Institute of Tropical Agriculture (IITA); 2012. 1-24. Available from: www.cta.int

[29] Oyewole OB. Cassava Processing in Africa. Washington DC: National Academy of Science; 1992. pp. 89-92

[30] Ojo O, Deane R. Effects of cassava processing methods on antinutritional components and health status of children. Journal Science Food Agriculture. 2002;**82**:3, 252-257

[31] Tunde-Akintunde TY, Olaniyan SA, Adelakun EO. Effect of drying method on selected properties of cassava flour (lafun). Lautech Journal Engineering Technology. 2007;**4**(2):19-23

[32] Padonou SW, Nielsen DS, Akissoe NH, Hounhouigan JD, Nago MC, Jakobsen M. Development of starter culture for improved processing of Lafun, an African fermented cassava food product. Journal Applied Microbiology. 2010;**109**(4):1402-1410

[33] Odunfa S, Oyewole O. African fermented foods. In: BJB W, editor. Microbiology of Fermented Foods. Boston: Springer; 1998. pp. 713-752

[34] Achi OK, Akomas NS. Comparative assessment of fermentation techniques in the processing of fufu , a traditional fermented cassava product. Pakistan Journal of Nutrition. 2006;**5**(3):224-229

[35] Umeh SO, Okeke BC, Achufusi JN, Emelugo BN. Dry fufu powder, an alternative method of fufu preservation for availability and easy transportation. International Journal of Agriculture and Biosciences. 2019;**8**(2):89-98

[36] Abass AB, Adegoke GO, Awoyale W, Gaspar A, Mlingi N, Andrianavalona V, et al. Enumeration of the microbiota and microbial metabolites in processed cassava products from Madagascar and Tanzania. Food Control. 2018;**99**:164-170. DOI: 10.1016/j.foodcont.2018.12.025

[37] Ellis WO, Dziedzoave NT, Boakye K, Simpson BK, Smith JP. Effect of cassava variety and processing methods on the performance of 'kudeme' in agbelima production. Food Control. 1997;**8**(4):199-204

[38] Dziedzoave NT, Ellis WO, Oldham JH, Oduro I. Optimizing agbelima production:varietal and fermentation effect on product quality. Food Research International. 2000;**33**:867-873

[39] Djeni NT, N'Guessan KF, Toka DM, Kouame KA, Dje KM. Quality of attieke (a fermented cassava product) from the three main processing zones in Côte d'Ivoire. Food Research International. 2011;**44**(1):410-416. DOI: 10.1016/j.foodres.2010.09.032

[40] Daouda N, Achille TF, Abodjo KC, Charlemagne N, Georges AN. Influence of traditional inoculum and fermentation time on the organoleptic quality of "attiéké.". Food Nutrition Science. 2012;**03**(10):1335-1339

[41] Bouatenin JKM, Djeni TN, Aka S, KBKM D. The contribution of microorganisms to the fermentation of cassava dough during attiéké processing in Côte d ' Ivoire. Food Glob Sci Books. 2012;**6**(1):58-64

[42] Ele IE. Technical Efficiency of Cassava Producers in Ikom Agricultural Zone of Cross River State. Nigeria Journal of Research in Agriculture and Animal Science. 2014;**2**(10):9-15

[43] Iwuagwu O. The spread of cassava (manioc) in Igboland , South-East Nigeria: A reappraisal of the evidence. The Agricultural History Review. 1985;**2012**:60-76

[44] Ifeabunike O, Nwaedozie J, Aghanwa C. Proximate analysis,

hydrogen cyanide and some essential mineral content of sweet cassava variety (*Manihot utilisima*) and bitter cassava variety (*Manihot palmata*) cultivated in Kachia Local Government area of Kaduna State, Nigeria. International Journal Biochemistry Research Review. 2017;**19**(1):1-12

[45] Ezedinma CI, Kormawa PM, Manyong VM, Dixon AGO. Challenges, opportunities, and strategy for cassava sub sector development in Nigeria. In: Proceedings of the 13th ISTRC Symposium. 2007. p. 627-640.

[46] Westby A. Cassava utilization, storage and small-scale processing. In: Hillocks RJ, Thresh JM, Bellotti AC, editors. Cassava: Biology, production and utilization. Wallingford: CABI Publishing; 2009. pp. 281-300. DOI: 10.1079/9780851995243.0281

[47] Balagopalan C. Cassava utilization in food, feed and industry. In: Hillocks RJ, Thresh JM editors. Cassava: Biology, Production and Utilization. Wallingford:CABI Publishing; 2001. p. 301-318

[48] African Union. Youth Development. 2021. Available from: https://au.int/en/youth-development [Accessed: 2 November 2021]

[49] Development Initiatives. Global Nutrition Report 2017: Nourishing the SDGs. Bristol UK Development Initiatives; 2017

[50] FAO, ECA and AUC. Africa Regional Overview of Food Security and Nutrition 2019. Accra: FAO, ECA and AUC; 2020. 10.4060/CA7343EN

[51] Statista, Share of undernourished people worldwide in 2020, by region 2022. Available from: https://www.statista.com/statistics/273291/number-of-people-with-malnutrition-worldwide/ [Accessed: 2022-03-14]

[52] Muthayya S, Rah JH, Sugimoto JD, Roos FF, Kraemer K, Black RE. The global hidden hunger indices and maps: an advocacy tool for action. PLoS One. 2013;**8**(6)

[53] UNECA. The Cost of Hunger in Africa: Social and Economic Impact of Child Undernutrition in Egypt, Ethiopia, Swaziland and Uganda. Report. Addis Ababa: African Union Commission, NEPAD Planning and Coordinating Agency. UN Economic Commission for Africa, and UN World Food Programme. 2014

[54] Kolapo AL, Sanni MO. A comparative evaluation of macronutrient and micronutrient profile of soybean-fortified gari and soybean fortified tapioca. Food and Nutrition Bulletin. 2009;**30**(1):90-94. DOI: 10.1177/156482650903000110

[55] Kolapo AL. Soybean: Africa's potential cinderella food crop. In: Ng T-B, editor. Soybean— Biochemistry, Chemistry and Physiology. Croatia: Intech; 2011. pp. 137-150 Available from: http://www.intechopen.com/articles/show/title/soybean-africa-s-potential-cinderella-food-crop

[56] Eufic. Evolution of Food Processing and Labelling in Food Production. 2014. Available from: https://www.eufic.org/en/food-production/article/food-production-1-3-the-evolution-of-meeting-nutritional-needs-through-proc [Accessed: 2022-03-17]

[57] Welch RW, Mitchell PC. Food processing: A century of change. British Medical Bulletin. 2000;**56**(1):1-17. DOI: 10.1258/0007142001902923

[58] Zweiniger-Bargielowska I, Duffett R, Drouard A, editors. Food and War in

20th Century Europe. England: Ashgate Publishing Limited; 2011

[59] Kolapo AL, Sanni MO. Processing and characteristics of soybean-fortified tapioca. Journal of Women in Technical Educations. 2005;**4**:36-44

[60] Aweke K. Detoxification and consumption of cassava-based foods in southwest Ethiopia. Pakistan Journal of Nutrition. 2012;**11**(3):237-242

[61] Adejumo BA, Raji AO. Microbial safety and sensory attributes of Gari in selected packaging materials. Natural and Applied Science. 2012;**3**:153-162

[62] Assanvo J, Agbo G, Behi Y. Microflora of traditional starter made from cassava for attiéké|| production in Dabou (Côte d'Ivoire). Food Control. 2006;**17**:37-41

[63] Adejuyitan JA, Otunola ET, Bolarinwa IF, Olanipekun BF. Changes in the protein and cyanide contents of pupuru as affected by duration of fermentation of cassava with species of *rhizopus*. Journal of Food Chemical Nanotechnology. 2017;**3**(1):19-23

[64] Aloys N, Zhou HM. Comparative study on nutrient and anti-nutrient changes in Ikivunde and Inyange, two Burundian traditionally processed cassava products. Journal of Science Food and Agriculture. 2006;**86**:1878-1886

[65] Bamidele OP, Fasogbon MB, Oladiran DA. Nutritional composition of fufu analog flour produced from Cassava root (*Manihot esculenta*) and Cocoyam (*Colocasia esculenta*) tuber. Food Science and Nutrition. 2015;**3**:597-603

[66] Halake NH, Chinthapalli B. Fermentation of traditional african cassava based foods: Microorganisms role in nutritional and safety value. Journal of Experimental Agriculture International. 2020;**42**(9):56-65

[67] Kolapo AL, Salami RA, Onipede GO. Molecular identification and technological properties of yeasts isolated from spontaneously fermented cassava waste pulp. Nova Biotechnologica *et* Chimica. 2021;**20**(2):e898

[68] Kolapo AL, Salami RA, Onajobi I, Oluwafemi F, Fawole AO, Adejumo OE. Detoxification and nutritional enrichment of cassava waste pulp using *Rhizopus oligosporos* and *Aspergillus niger* The Annal. Universit. Dunarea de Jos of Galati Fascicle VI – Food Technology. 2021;**45**:52-68

[69] Fawole AO, Kolapo AL, Omotoso SK, Onipede GO, Bolarinwa OO. Nutritional enrichment of *Lafun*: An African fermented cassava food, using *Lactobacillus plantarum* and *Saccharomyces boulardii*. International Journal of Food Science and Nutrition. 2021;**6**(2):55-59

[70] Oguntoyinbo FA. Identification and functional properties of dominant lactic acid bacteria isolated at different stages of solid state fermentation of cassava during traditional gari production. World Journal of Microbiology and Biotechnology. 2007;**23**(10):1425-1432. DOI: 10.1007/s11274-007-9386-7

[71] Sobowale AO, Olurin TO, Oyewole OB. Effect of lactic acid bacteria starter culture fermentation of cassava on chemical and sensory characteristics of fufu flour. African Journal of Biotechnology. 2007;**6**(16):1954-1958

[72] Bouatenin KMJ, Djeni NT, Kouame KA, Charles Y, Bi T, Herve E. Use of starter cultures of lactic acid bacteria , yeasts , bacilli and moulds in the fermentation of cassava dough for attieke (an ivoirian fermented food) preparation. International Journal of Innovation and Applied Studies. 2017;**21**(1):48-58

[73] Amoa-Awua WKA, Appoh FE, Jakobsen M. Lactic acid fermentation of cassava dough into agbelima. International Journal Food Microbiology. 1996;**31**:87-98

[74] Verce M, De Vuyst L, Weckx S. Shotgun metagenomics of a water kefir fermentation ecosystem reveals a novel *Oenococcus* species. Frontiers in Microbiology. 2019;**10**:479

[75] Ray RC, Joshi VK. Fermented foods: Past, present and future. In: Ray RC, Didier M, editors. Microorganisms and Fermentation of Traditional Foods. 1st ed. Boca Raton: CRC Press; 2014. DOI: 10.1201/b17307

[76] Obafemi YD, Oranusi SU, Ajanaku KO, Akinduti PA, Leech J, Cotter PD. African fermented foods: overview, emerging benefits, and novel approaches to microbiome profiling. NPJ Science of Food. 2022;**6**(15). DOI: 10.1038/s41538-022-00130-w

[77] Walsh AM, Crispie F, Daari K, O'Sullivan O, Martin JC, Arthur CT, et al. Strain-level metagenomic analysis of the fermented dairy beverage nunu highlights potential food safety risks. Applied Environmental. Microbiology. 2017;**83**:e01144-e01117

[78] Jain C, Rodriguez-R LM, Phillippy AM, Konstantinidis KT, Huru S. High throughput ANI analysis of 90K prokaryotic genomes reveals clear species boundaries. Nature Communications. 2018;**9**:1-8. DOI: 10.1038/s41467-018-07641-9

[79] Leech J, Cabrera-Rubio R, Walsh AM, Macori G, Walsh CJ, Barton W, et al. Fermented-food metagenomics reveals substrate-associated differences in taxonomy and health-associated and antibiotic resistance determinants. Applied and Environmental Science. 2020;**5**(6):e00522-e00520. DOI: 10.1128/mSystems.00522-20

[80] Brenner K, You L, Arnold FH. Engineering microbial consortia: a new frontier in synthetic biology. Trends in Biotechnology. 2008;**26**:483-489

[81] van HylckamaVlieg JE, Veiga P, Zhang C, Derrien M, Zhao L. Impact of microbial transformation of food on health—From fermented foods to fermentation in the gastrointestinal tract. Current Opinion in Biotechnology. 2011;**22**:1-9

[82] Hugenschmidt S, Miescher Schwenninger S, Lacroix C. Concurrent high production of natural folate and vitamin B12 using a co-culture process with *Lactobacillus plantarum* SM39 and *Propionibacterium freudenreichii* DF13. Process Biochemistry. 2011;**46**:1063-1070

[83] Hugenschmidt S, Miescher Schwenninger S, Gnehm N, Lacroix C. Screening of a natural biodiversity of lactic and propionic acid bacteria for folate and vitamin B12 production in supplemented whey permeate. International Dairy Journal. 2010;**20**:852-857

[84] Kolapo AL, Popoola TOS, Afolabi OR, Atanda OO, Oluwafemi F. Evaluation of spontaneously fermenting soybean daddawa microbiota's potential for starter culture application. International Journal of Research and Innovation in Applied Science. 2019;**4**(3):41-48

[85] Kolapo AL, Popoola TOS, Omafuvbe OB. Biochemical characterization of multiple starter cultures fermentation of soybean (*glycine max* l.) seeds for soybean daddawa production annals. Food Science and Technology. 2019;**20**:688-700

[86] Bartkiene E, Juodeikiene G, Vidmantiene D, Viskelis P, Urbonaviciene D. Nutritional and quality aspects of wheat sourdough bread using *Lactobacillus luteus* and *L. angustifolius* flours fermented by *Pediococcus acidilactici*. International Journal of Food Science and Technology. 2011;**46**:1724-1733

[87] Holzapfel WH. Appropriate starter culture technologies for small-scale fermentation in developing countries. International Journal of Food Microbiology. 2002;**75**:197-212

[88] Moroni AV, Dal Bello F, Arendt EK. Sourdough in gluten-free bread-making: an ancient technology to solve a novel issue? Food Microbiology. 2009;**26**:676-684